Imagine Listening

Your worst day is our everyday...

RICARDO MARTINEZ II

WITHIN THE TRENCHES MEDIA

code seven traffic, llc

Imagine Listening
Your worst day is our everyday Vol. 2

Copyright © 2024 Ricardo Martinez, II.
All rights reserved.

No portion of this book may be reproduced in whole or in part, by any means whatsoever, except for passages excerpted for the purposes of review, without the prior written permission of the publisher. For information or to order additional copies, please contact:

Code Seven Traffic, LLC
828 Mannes Pine Cove Fort Wayne, IN 46814
260.402.1810 | withinthetrenches.net
Publisher: Ricardo Martinez II
Editor: Ricardo Martinez II

PUBLISHER'S CATALOGING-IN-PUBLICATION DATA:

Name: Martinez II, Ricardo, author | Title: Imagine Listening: Your worst day is our everyday Vol. 2 / Martinez II, Ricardo Description: Fort Wayne, IN: Code Seven Traffic, [2024]

Identifiers: ISBN: 979-8218467821

Subjects: LCSH: Martinez II, Ricardo, author | Fire & Emergency Services: Biography & Autobiography | Criminology: Social
| Criminology: Social Science | Career - Essay | Public Safety Telecommunications -- Personal narratives. | Career: Public Safety, 911. | Autobiographies. | BISAC: BIOGRAPHY & AUTOBIOGRAPHY / Fire & Emergency Services
/ SOCIAL SCIENCES / Criminology | Career | Essay.

Acknowledgments

I would like to take the opportunity to thank a couple of people for playing pivotal roles in my public safety journey. Years ago I walked into Frostproof Police Department in Frostproof, Florida. A few days prior, I had applied for the position of 9-1-1 dispatcher. I didn't expect a call back as I had no prior experience. The only experience I had with 9-1-1 was a TV show hosted by William Shatner called, "Rescue 911."

I was scared. The records clerk, Candi, opened the door from the lobby where I was sitting and escorted me a few steps down to the Chief's office. I walked in and met Chief Byrd and Lieutenant Ard. Once the interview began my nerves went away. My fear was gone and when the interview was over I was excited. After two months of waiting, I was hired and began my public safety journey. Thank you, Chief Byrd, for taking a chance on someone whose experience in 9-1-1 was based on a show he grew up with. I will be forever grateful to you.

Fast forward several years to the continuation of my dispatch career in Michigan. I was given an opportunity to work in the county I grew up in. To my surprise, I also ended up working with a childhood friend of mine, Megan. I was excited to work with her and many others, but while in training, things were rough. I almost quit. I remember pulling Megan aside and I told her I was thinking of quitting because the training process was a bit lonely. I thought we were supposed to be a team, yet I felt alone with so many people around me. She took the time to listen to me and in the end, she told me to stick it out, that everything would be ok. Thank you, Megan, for taking the time to hear me out. It's because of you that I stayed. Had it not been for you, everything I have done in 9-1-1, might not exist.

As always, I want to thank the entire Thin Gold Line. It is because of you that I can do what I do. Your support and trust in me to share your stories is something I cherish and hold close to my heart. It is because of you that I do what I do. You are my WHY.

Warning: The stories you are about to read are all true and come from the 9-1-1 professionals who have lived them. If you have PTSD or suicidal tendencies, you are reading at your own risk.

I REMEMBER THE SCREAMS

I REMEMBER YOUR LAST BREATHS

I THINK OF YOU OFTEN, I REMEMBER YOU

Welcome to *Imagine Listening - Your worst day is our everyday*. This is book two in a series of books that will feature stories from the #IAM911 Movement.

Imagine listening
Your worst day is our everyday

foreword

Most people don't realize there are four pillars to public safety. The first pillar is the Emergency Communications Center (ECC), followed by Law Enforcement, Fire, and Emergency Medical Services (EMS). That first pillar goes unseen but is so crucial that those experiencing an emergency would be lost without it.

An emergency response generally starts with dialing the three most essential numbers everyone should know—9-1-1. The 9-1-1 professionals, AKA Public Safety Telecommunicators (PSTs), who answer those calls have long been the "unsung heroes" of the public safety community. They work tirelessly behind the scenes of every emergency and non-emergency call, virtually holding the scene down until field responders arrive and coordinating the public safety response. Those calling for help often believe they are speaking directly with the police, fire, or EMS responders en route to their distress call rather than those unseen heroes in the ECC.

I spent 24 years in an ECC. Throughout my career, I saved many lives, coordinated countless emergency responses, and investigated and gathered data that allowed law enforcement to respond to and mitigate countless crimes in progress. Anyone working in an ECC understands they won't likely get credit for that work or even receive a thank-you. And to be fair, they aren't in it for the glory. They are there because they have a drive to help others.

The high volume of vicarious trauma experienced by our profession are profound, which leads to high rates of burnout and post-traumatic stress injuries. Over the last 10-15 years, public safety responders' mental health and wellness have come to the forefront, but PSTs were not included in that. Unfortunately, at the federal level, PSTs are classified as Administrative/Clerical employees. Although these unsung heroes handle higher volume of vicarious trauma than their other public safety brethren, they don't qualify for those benefits because they don't fall under the right job classification.

NENA: The 9-1-1 Association and the Association of Public Safety Officials (APCO) began working to educate the US Bureau of Labor and Statistics (BLS) before their last 10-year review of the PST job classification. Que, my dear friend, and brother from another mother, Ricardo Martinez. His call to arms helped the greater 9-1-1 community lend their voice to this important cause. Initially, the movement was meant to tell the story of these vital unsung heroes to the Bureau of Labor and Statistics, but it quickly became about much more than that. It became an outlet for the mental health and wellness of our 9-1-1 professionals. Telling their stories has brought this profession to the forefront rather than being in the shadows. It has allowed those who have dealt with so many tragedies to be acknowledged and to recognize that it's okay not to be ok sometimes. Their voices matter because of the #IAM911 movement. The 100,000 plus 9-1-1 professionals who make a difference every day deserve to be recognized as the vital link within the public safety community that they are. I will forever be grateful that Ricardo chose to tell our story!

April Heinze, ENP
Chief of 9-1-1 Operations,
NENA: The 9-1-1 Association

#IAM911

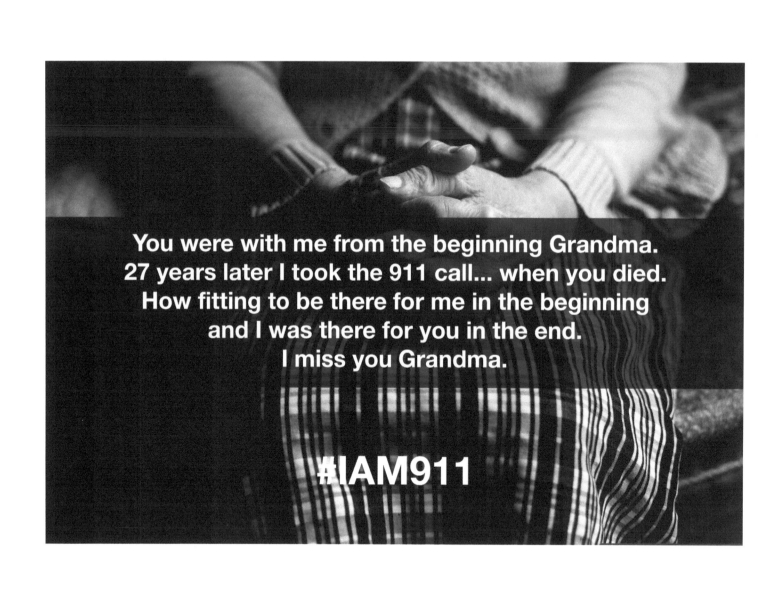

Introduction

John Lennon and Paul McCartney wrote,

"Oh, I get by with a little help from my friends
Mm, I get high with a little help from my friends
Mm, gonna try with a little help from my friends"

Many things we do in life come with a little help from our friends and when it comes to the birth of the #IAM911 Movement, it holds true. In the beginning, stories flooded in from all over the world. They continue to do so, but not as frequently as it was early on. I added four administrators to my team to tackle the stories, but I quickly added two more. I remember sitting on my couch with my laptop late at night. I would read a story from the Within the Trenches podcast page inbox and as I read through it I would see several more stories pop up. It was INSANE!

In our case, the "high" with a little help from my friends revolved around the feeling of helping so many of our sisters and brothers share stories that they had buried for years. I remember feeling a massive weight lift off my shoulders as each story went out. The weight was not my own, but the weight of the person who had shared their story.

One person said, "My hands are still shaking from typing out my story but I feel so much better. Thank you."

Others over the years have said something similar. They are happy to have an outlet to share their stories with the world. The success of Imagine Listening wouldn't be what it is today if not for the amount of stories that were pushed out as fast as they came in back then. For that, I want to thank my original six. Thank you Daphanie, Jessica, Jordon, Kelli, Victoria, and Shae. "Oh, I get by with a little help from my friends."

A few months after the launch of book one of Imagine Listening, I received a couple of emails concerning the #IAM911 stories in the book. People were looking for more. What took place after the story? Did the person live? I have a family member who is a 911 dispatcher and there is more to the story, but it wasn't added. How can this be?

While each question is valid, I want to point out two things. Each story is written in a way that leaves you wondering what happened. The reason for this is that one of the things we deal with as 911 professionals is closure. We don't always know what the outcome of a call is. There will be times that we receive the gift of closure but the majority of the time we are in the dark. Each story is meant for you to experience what it is like for us.

The second point I want to make clear is that every story was posted with the amount of information that each 911 professional wanted to have out there. Because I have worked and experienced similar calls, I respected their wishes. I knew what they meant. I knew what they felt and only published what they wanted. This is something that I will continue to do for as long as the #IAM911 Movement and Imagine Listening is around.

With that said, you will get the chance to learn more about a few stories; stories of my own. They are not ones that I talk about often, but it is therapeutic for me to share them. And by sharing, I hope that it will inspire others to share their stories when they are ready to do so. In the following pages, you will continue to learn from 9-1-1 professionals through the power of storytelling. While some stories have a good outcome, the majority are horrifying. This is their reality. Your worst day is their every day.

So, let's start with a bang. I want you to imagine each story. Become the dispatcher who has taken each call and imagine listening…

Imagine Listening

Your worst day is our everyday...

Stories

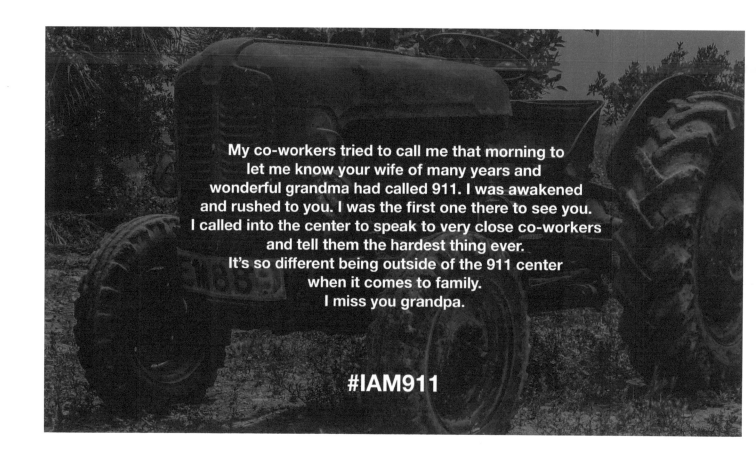

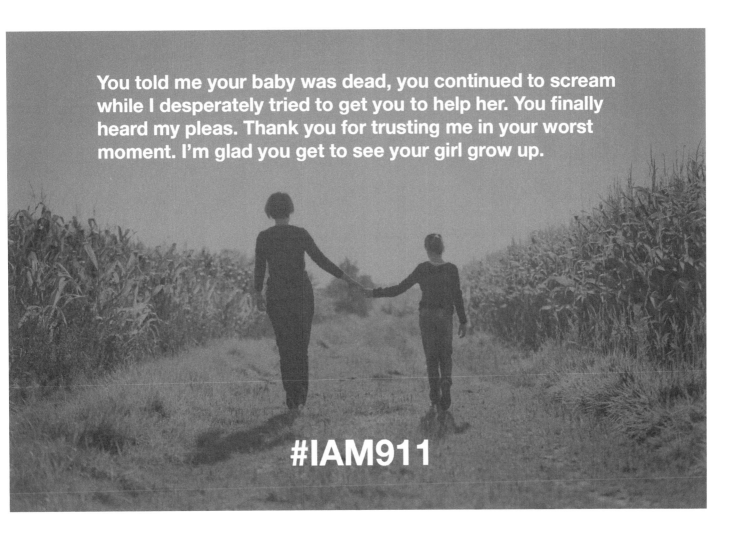

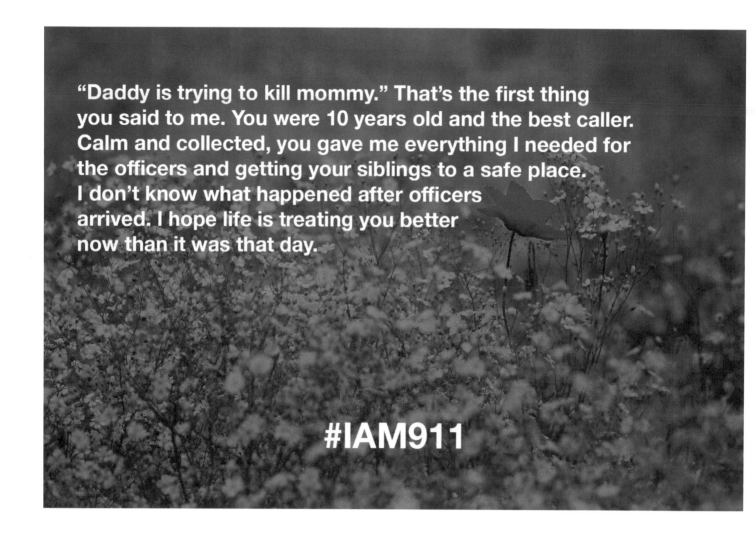

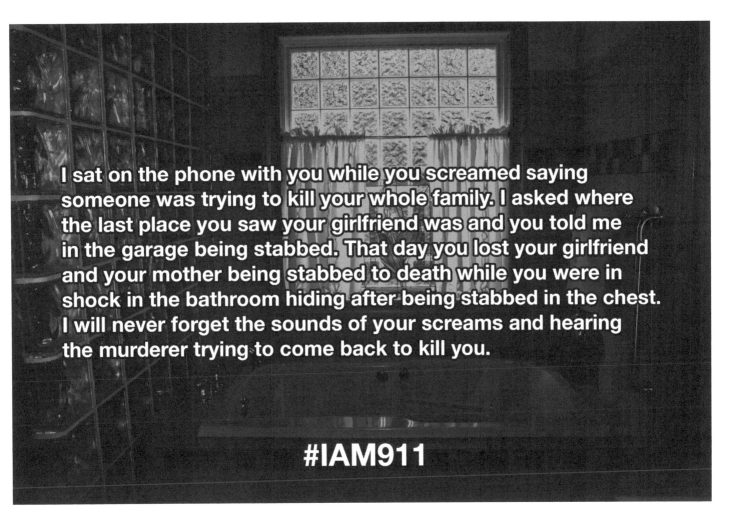

I sat on the phone with you while you screamed saying someone was trying to kill your whole family. I asked where the last place you saw your girlfriend was and you told me in the garage being stabbed. That day you lost your girlfriend and your mother being stabbed to death while you were in shock in the bathroom hiding after being stabbed in the chest. I will never forget the sounds of your screams and hearing the murderer trying to come back to kill you.

#IAM911

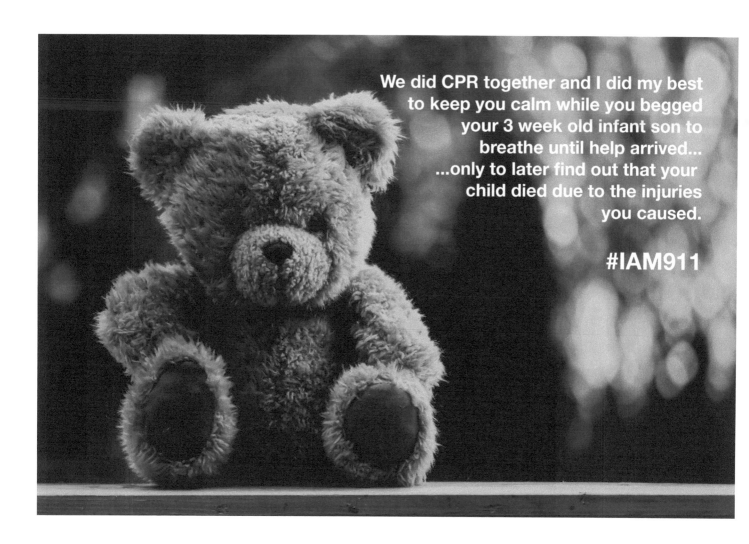

My husband told me it was the most gruesome scene he had been on in his 8 years of being an officer. After disconnecting the line I got up and went outside.
I cried for you.
That was 10 yrs ago and you are still with me to this day.

#IAM911

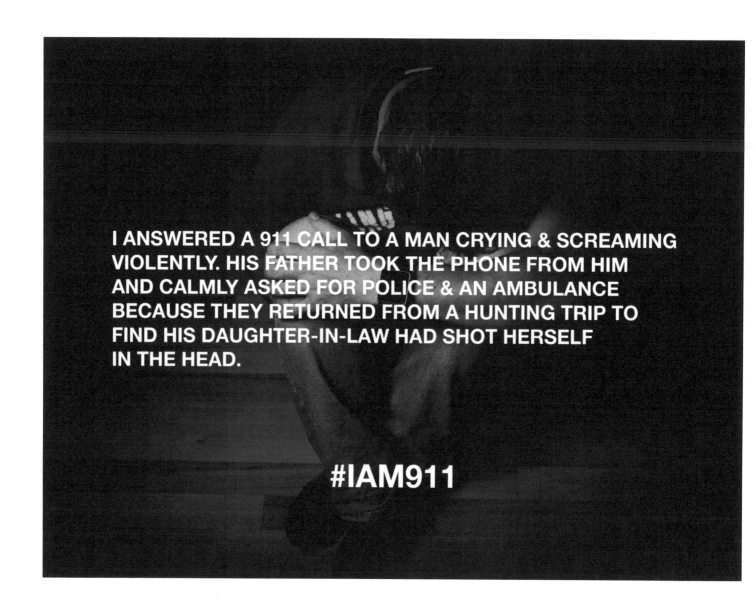

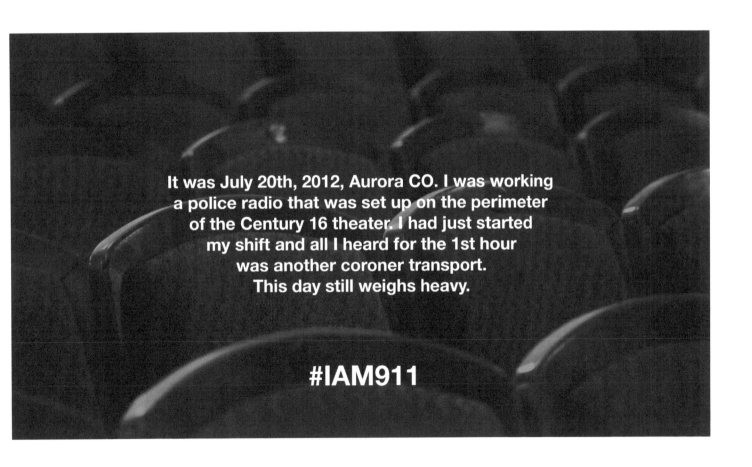

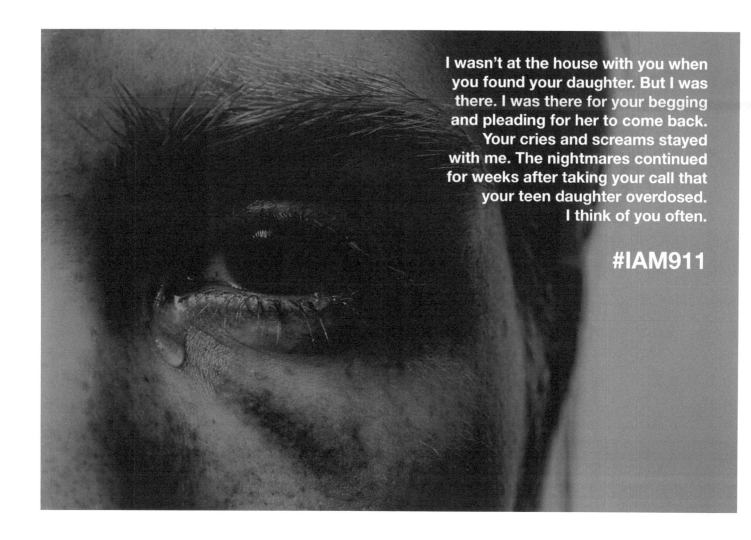

I was at a funeral for a friend's son that had committed suicide. While at the dinner, afterwards, I was speaking to the mother of my friend I had just met. A lady approached her and introduced herself. It was the 911 dispatcher who took her 911 call.
They hugged and held onto each other. The dispatcher just wanted to tell her that she did a good job.

What a great reminder that we do make a difference.

#IAM911

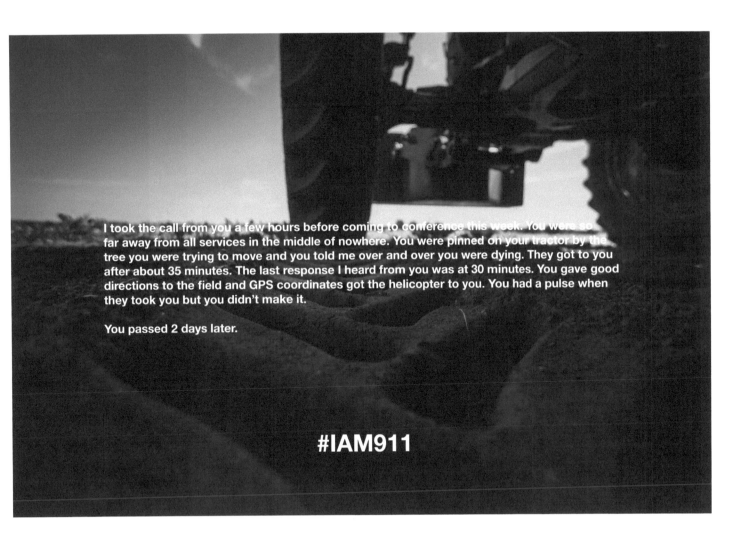

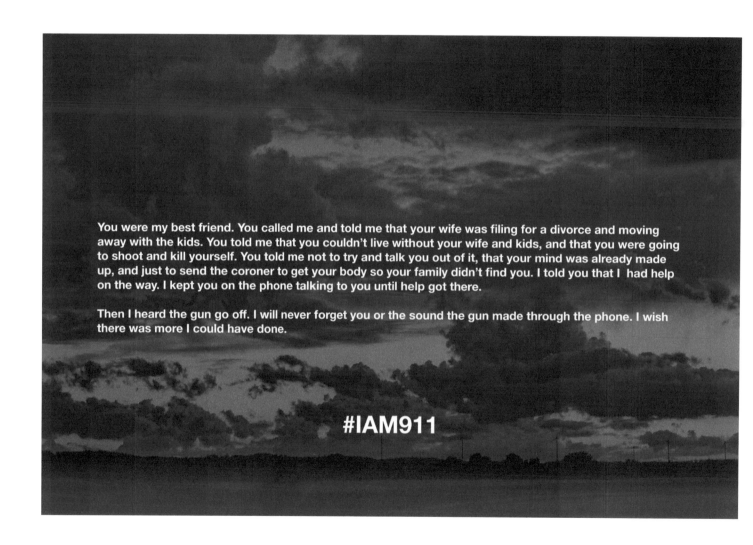

You were my best friend. You called me and told me that your wife was filing for a divorce and moving away with the kids. You told me that you couldn't live without your wife and kids, and that you were going to shoot and kill yourself. You told me not to try and talk you out of it, that your mind was already made up, and just to send the coroner to get your body so your family didn't find you. I told you that I had help on the way. I kept you on the phone talking to you until help got there.

Then I heard the gun go off. I will never forget you or the sound the gun made through the phone. I wish there was more I could have done.

#IAM911

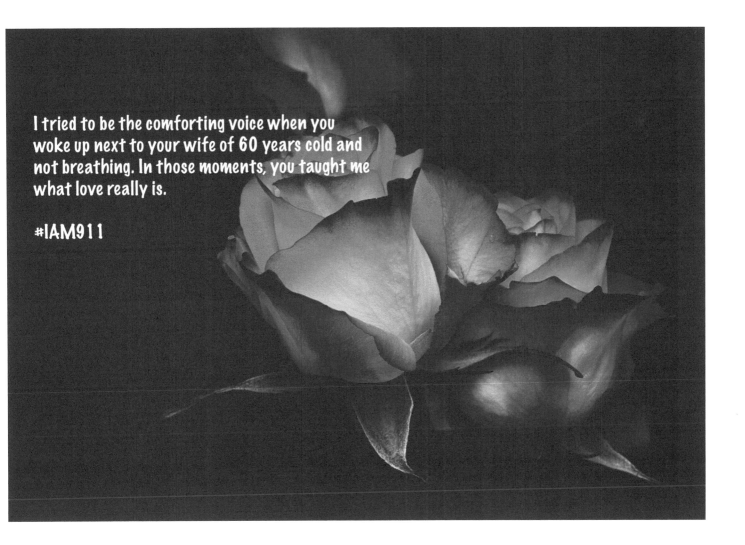

I never met you. I have never heard your voice. I met your husband's voice that day he called, frantic to save you. We tried to save your life together that day, we didn't succeed. We lost you together that day. You were my first working CPR call. His voice will forever stay with me.

I still hear him. I will never forget you, even though I never met you.

#IAM911

I wasn't dispatching the night you got shot,
but I feel like I should've been there.
It still haunts me that we almost lost you.

#IAM911

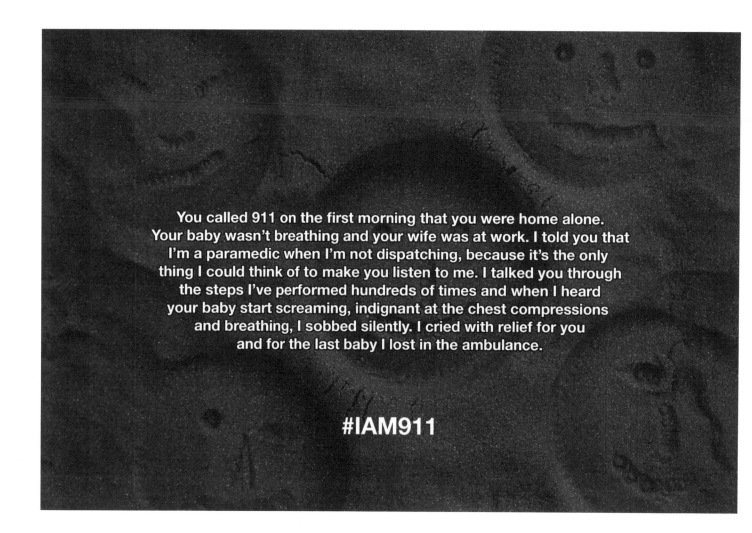

You called 911 on the first morning that you were home alone. Your baby wasn't breathing and your wife was at work. I told you that I'm a paramedic when I'm not dispatching, because it's the only thing I could think of to make you listen to me. I talked you through the steps I've performed hundreds of times and when I heard your baby start screaming, indignant at the chest compressions and breathing, I sobbed silently. I cried with relief for you and for the last baby I lost in the ambulance.

#IAM911

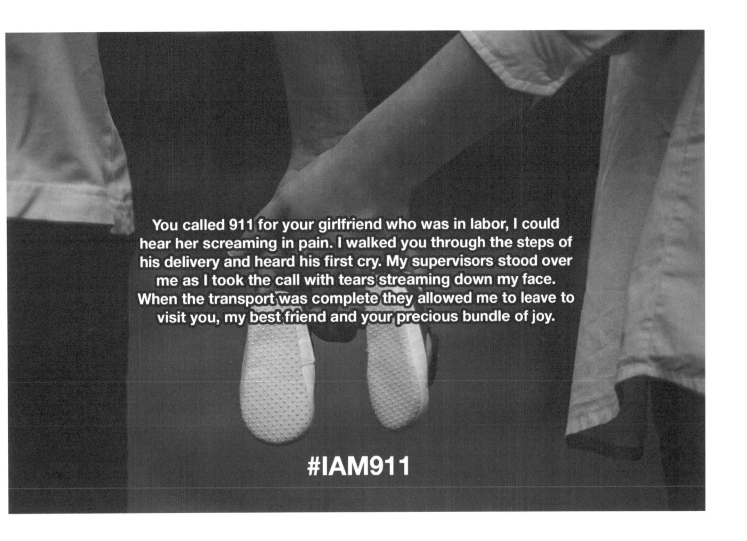

I answered your frantic call when you awoke to find your husband not breathing. I helped you do CPR on him until help arrived. It put a smile on my face to hear that he had a pulse on the way to the hospital.

#IAM911

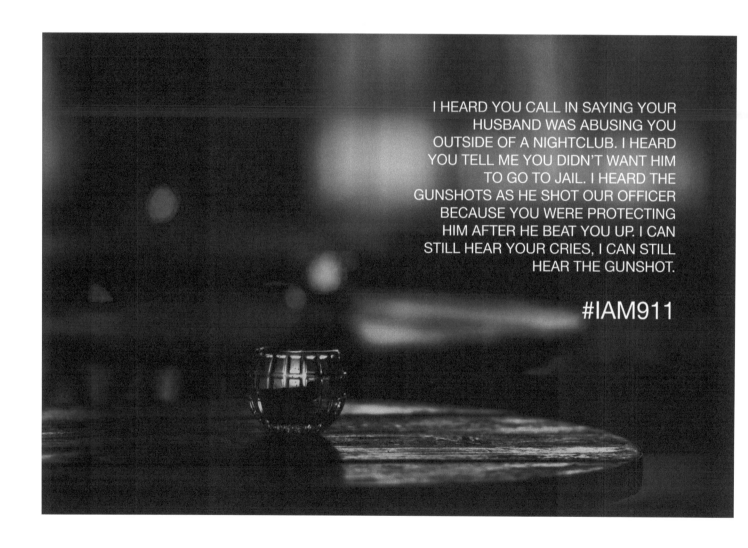

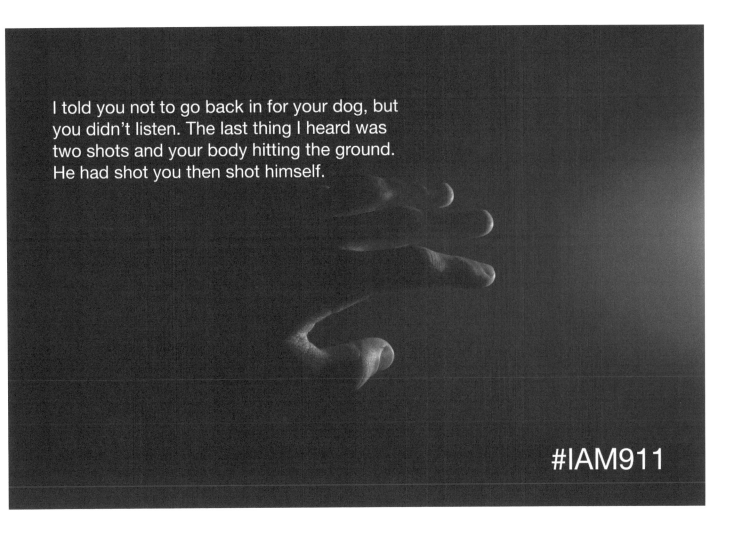

YOU REPORTED A BREAK IN.
SAID SOMEBODY KILLED HER...
MY HEART BROKE FOR YOU.
LATER YOU CONFESSED.
I'M GLAD YOU WERE FOUND GUILTY.
MY HEART IS STILL BROKEN.
SHE WAS 5 YEARS OLD.

#IAM911

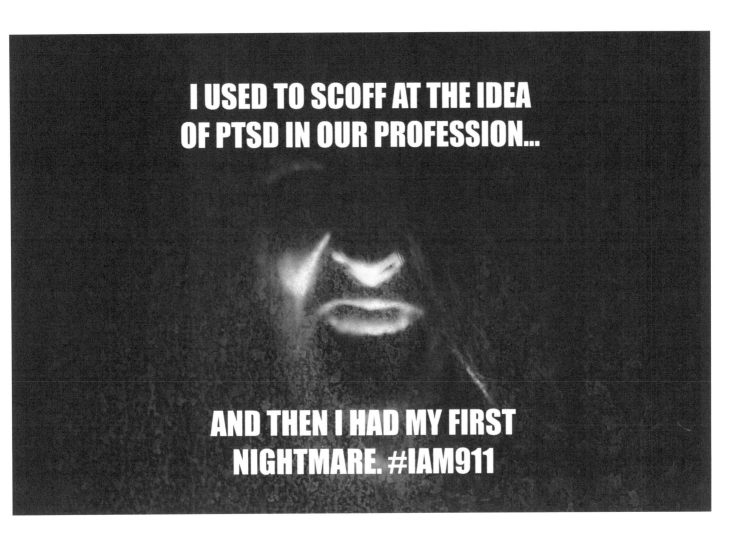

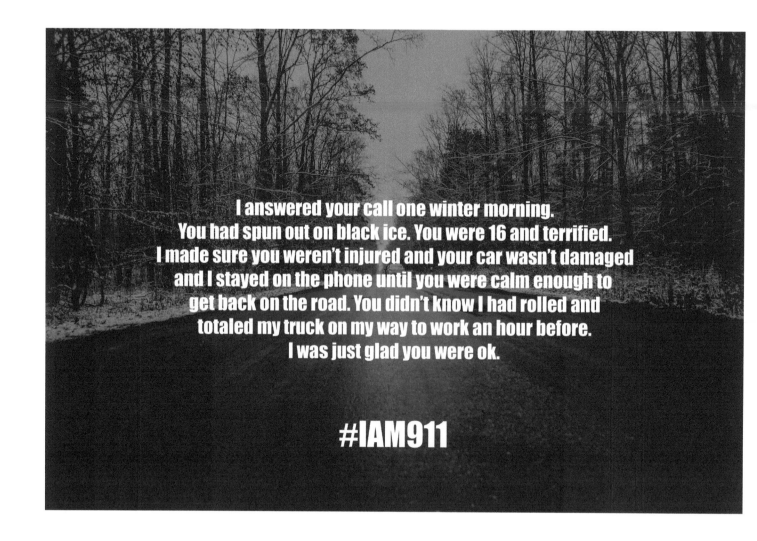

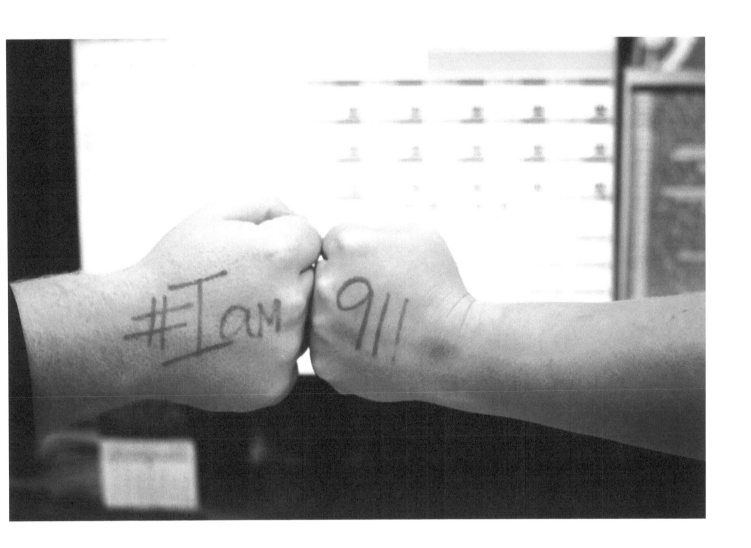

I knew her

Throughout our lives, we come in contact with people who leave an impression on us. Some of the people we come in contact with are meant to teach us a lesson or bring happiness, and some are single-serving, but others become lifelong friends. It's all about the connections we make and hold close to our hearts.

And then there are some people who make an impression and are suddenly gone. Those are the ones that sometimes hurt the most. Over twenty years ago, when I first started dispatching, I would go to a local restaurant with my mother and cousin. Every morning that I wasn't dispatching I was there with them having breakfast. It became our thing.

The food was always great, but the best part for me was our waitress. For privacy reasons I will make up a name for her. Joan was in her early 40's, shoulder length curly brown hair, average height and build, and she always had a smile on her face. I think what I loved the most about Joan was her sense of humor. We would sometimes chat with her for a while after she took our order and she would say,

"I should probably see if any of the other customers need something."

She would laugh hard and walk off to check on everyone else. I would look forward to seeing her when my family and I met up for breakfast because she was so full of energy. She had a love for life that rubbed off and you would be ready to take on the world. These are the types of people that we need in our lives. That energy and vibe is contagious. We continued to see her day after day, month after month, but after a while, we saw less and less of her. At one point we noticed that she was gone. She no longer worked at this restaurant and I had hoped that one day I would see her again.

It was an early morning in dispatch. I was drinking coffee and something felt off.

The phone rings, "911, where is your emergency?"

"I NEED HELP! PLEASE SEND SOMEONE!"

I listened as my caller frantically tried to tell me his location.

"Sir, what's going on?"

"It's my girlfriend!"

Imagine, for a moment, sitting on the couch with your loved one as you watch your favorite show. You start to doze off and you get up so that you can head to bed. Your loved one decides to finish the show and you kiss them goodnight. You climb into bed, lay your head on your pillow, close your eyes, and drift off.

This is what happened to my caller. There was nothing out of the ordinary. It was just another night, but when he woke up she wasn't there. He called out to her, but she didn't answer him. He called again, nothing. No response. My caller started to worry a little and got up. He walked into the living room and found his girlfriend still on the couch.

"I figured she was still asleep but when I checked on her she was blue and cold."

My heart broke for him as he told me what was going on.

"Sir, I have fire rescue coming your way already, but I need to transfer you to the county for EMS."

"Ok. Thank you."

I listened as he continued to mourn the loss of his girlfriend.

This entire call was less than a minute, but a ton of information was given during that time. Where I worked, at this small police department, we didn't provide medical instructions. We would take the initial call, attain the pertinent information and transfer the call to what we called, "County Radio" for EMS. We would send out Fire Rescue to assist EMS and an officer would respond as well for some calls.

After the call my heart sank. How would I feel if something like this happened to me? How would my caller go on after dealing with this? How would this ultimately impact him? These are questions that we as 911 professionals sometimes ask ourselves. I sat in silence as not many calls were coming in.

A few hours later the officer who had responded to the scene, returned. He stood next to me and told me what happened. This was the first time that I was given the gift of closure. As I have mentioned before, closure is one thing that we as 911 professionals struggle with. It is something that we don't always receive. We are left wondering what happened. There have been times, for myself personally, and I'm sure for others where I felt like maybe it was better for me to have stayed in the dark.

My officer told me that it was possibly an accidental overdose of prescription medications. What will always stick with me is that when my officer was telling me about my caller's girlfriend he mentioned that she had been working as a waitress for a long time but had recently left. As he described her my heart sank. The feeling I had all morning started to make sense. My heart was now breaking. I held my breath as my officer told me that he had pictures from the scene. I asked to see them.

"Do you know her?"

"Please let me see them," I asked.

"Are you sure," my officer asked.

I needed to confirm whether or not this was my friend who I had not seen at the restaurant in so long. In my head I prayed that it wasn't her. I took the photos from my officer as he handed them to me and I gasped. It was Joan.

"I know her, man. She was my favorite waitress at the restaurant."

"I'm sorry…"

Anything can happen in 911. When you work in the area that you live in it's only a matter of time before you take calls for or from people that you know. They could be friends, they could be family, and it's a hard thing to do. This was over twenty years ago, but it feels like I just took this call yesterday. What I mean is that the images of that day are in my head and in detail. Not just the photos that I saw, but that morning. The call coming in, what I heard, what I pictured in my head while my caller explained what he was seeing, and my conversation with the officer that responded to the call.

It's therapeutic to talk about my calls, but what I like to focus on is the constant smile and laughter that I enjoyed with Joan. Thank you for being a shining light in a world that can be dark and cold at times. Rest in peace, my friend.

Imagine Listening

Vol. 2

Your worst day is our everyday...

Stories

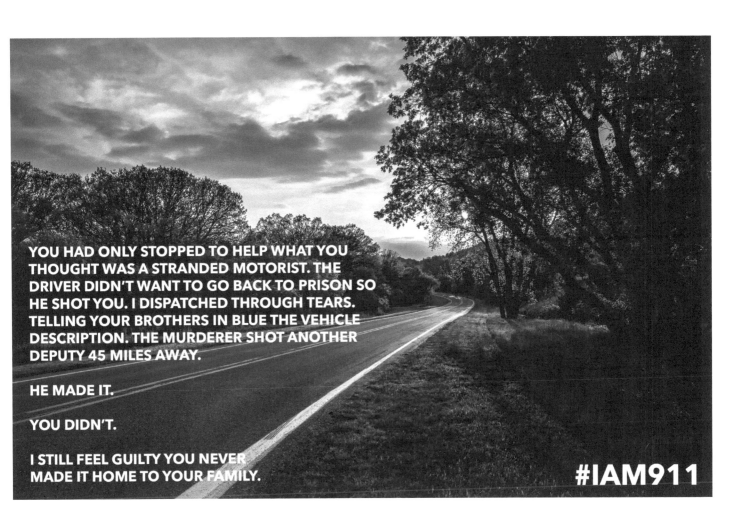

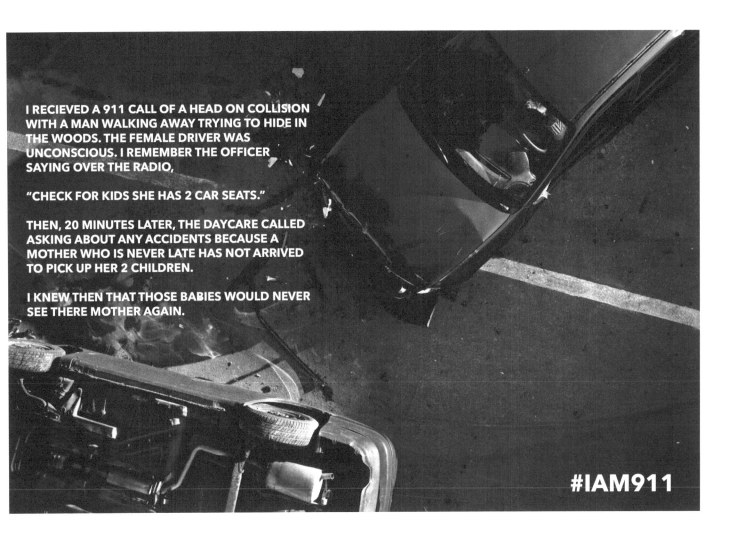

I RECIEVED A 911 CALL OF A HEAD ON COLLISION WITH A MAN WALKING AWAY TRYING TO HIDE IN THE WOODS. THE FEMALE DRIVER WAS UNCONSCIOUS. I REMEMBER THE OFFICER SAYING OVER THE RADIO,

"CHECK FOR KIDS SHE HAS 2 CAR SEATS."

THEN, 20 MINUTES LATER, THE DAYCARE CALLED ASKING ABOUT ANY ACCIDENTS BECAUSE A MOTHER WHO IS NEVER LATE HAS NOT ARRIVED TO PICK UP HER 2 CHILDREN.

I KNEW THEN THAT THOSE BABIES WOULD NEVER SEE THERE MOTHER AGAIN.

#IAM911

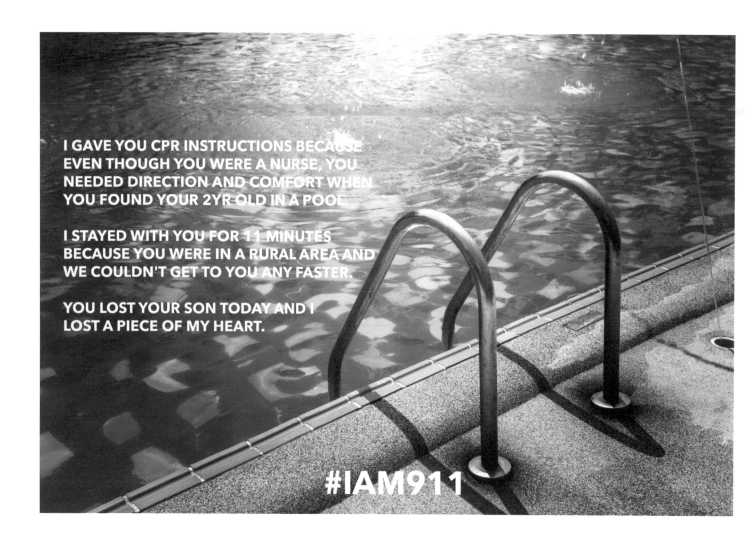

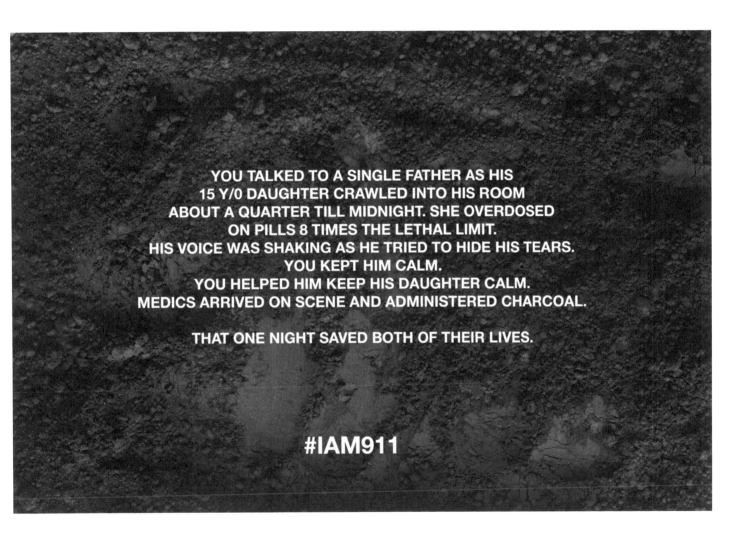

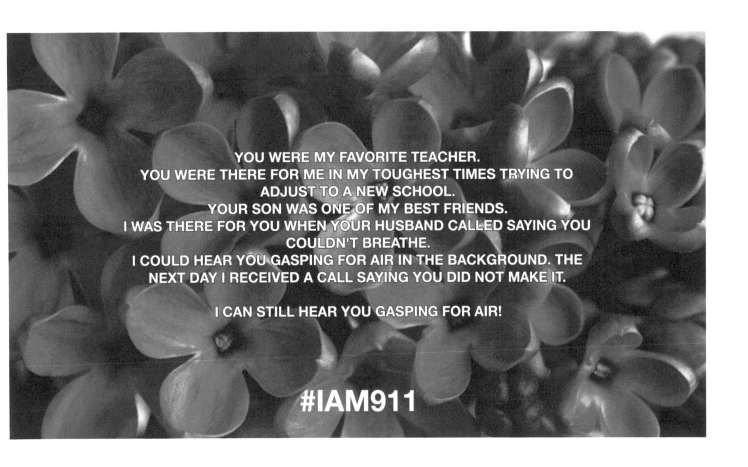

YOURS WAS THE FIRST FATAL ACCIDENT I EVER ATTENDED. I STILL REMEMBER HELPING LAY A WHITE BLANKET ACROSS TWO SEATS. I STILL REMEMBER BROADCASTING COMMS THAT YOU WERE DECEASED.

I STILL REMEMBER WATCHING YOUR GIRLFRIEND PACING THE ROADSIDE.

14 YEARS LATER I STILL REMEMBER...

#IAM000

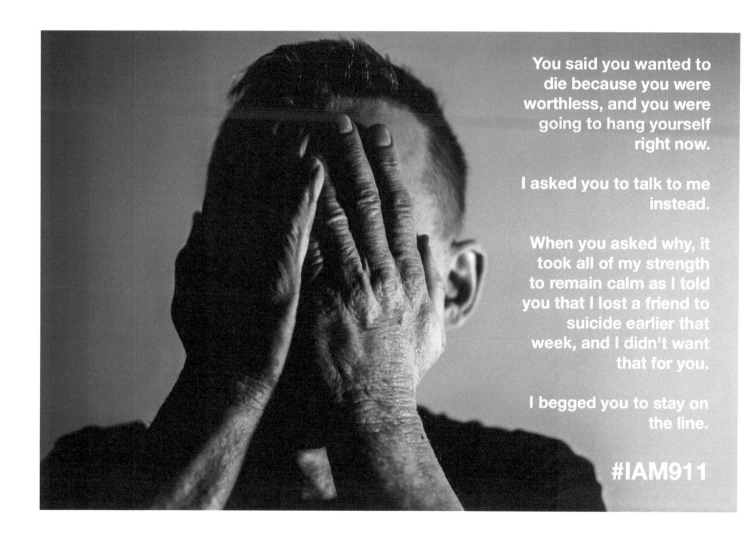

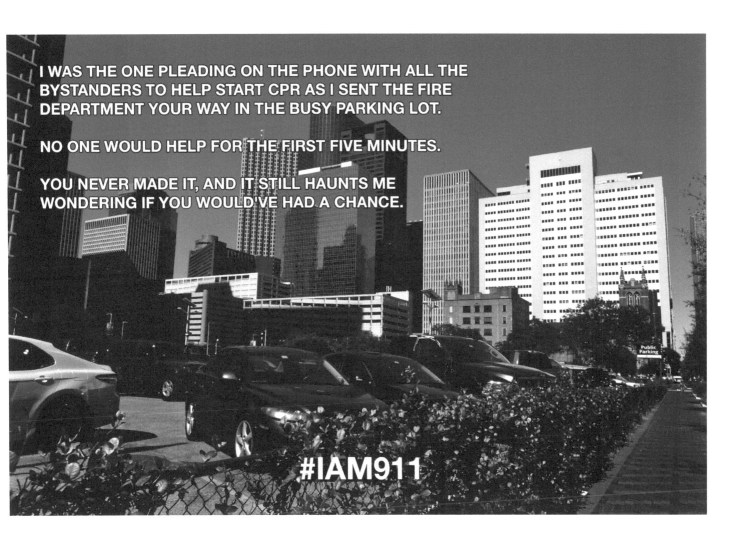

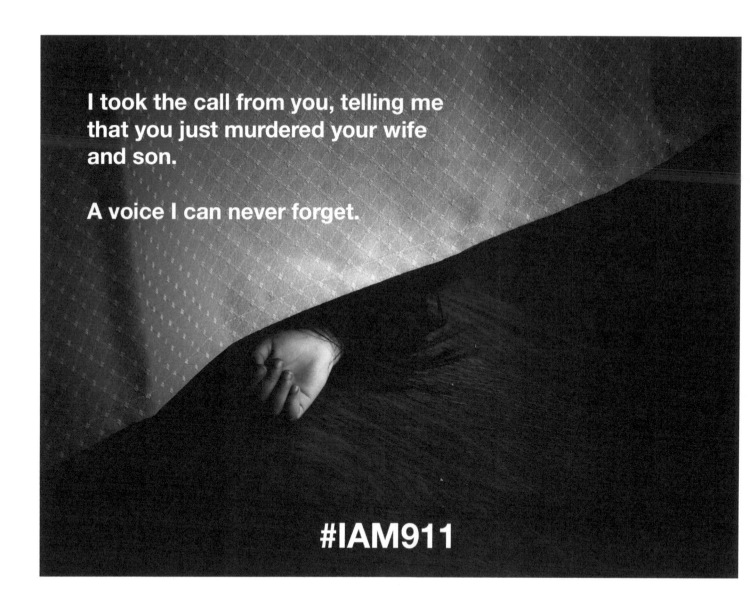

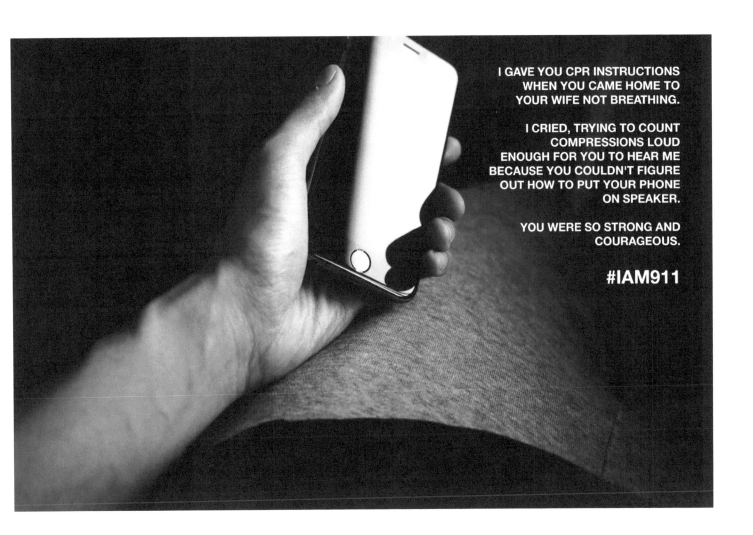

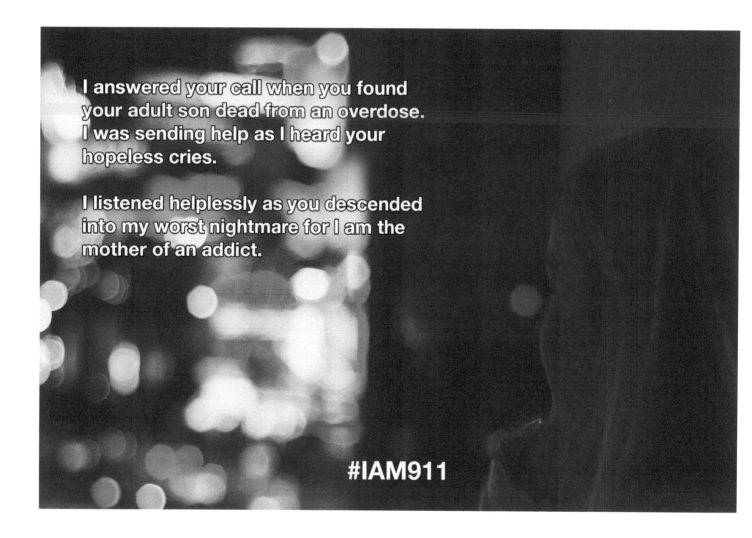

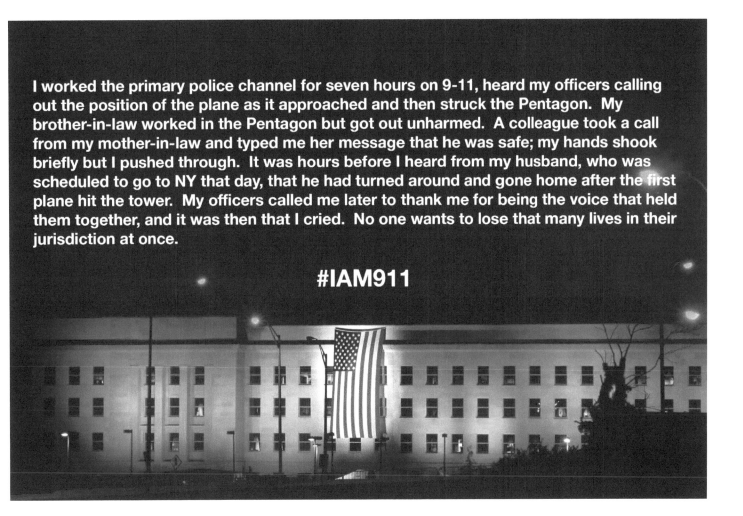

I worked the primary police channel for seven hours on 9-11, heard my officers calling out the position of the plane as it approached and then struck the Pentagon. My brother-in-law worked in the Pentagon but got out unharmed. A colleague took a call from my mother-in-law and typed me her message that he was safe; my hands shook briefly but I pushed through. It was hours before I heard from my husband, who was scheduled to go to NY that day, that he had turned around and gone home after the first plane hit the tower. My officers called me later to thank me for being the voice that held them together, and it was then that I cried. No one wants to lose that many lives in their jurisdiction at once.

#IAM911

#VEGASSTRONG

You were running away from the concert, you made your way to the parking lot of the church just to the east. You called because you came across two gunshot victims, both beyond help. I told you to just keep moving because there was nothing we can do and I had many other calls to get to but I heard something in your voice. That crack. That crack of someone trying to hold it together, but your control was slipping away. That crack in your voice let me know that while you might not be physically injured, you were most definitely mentally and emotionally injured. I talked to you for a few more minutes while you got further away to safety. While I know you will never be counted as one of the many injured, you most definitely were, and I will never ever forget that crack.

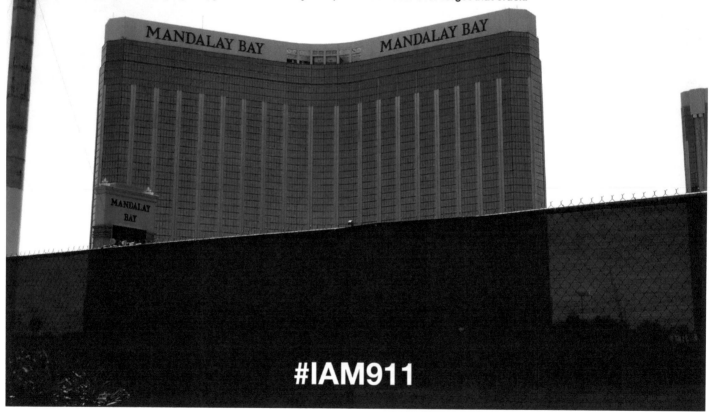

#IAM911

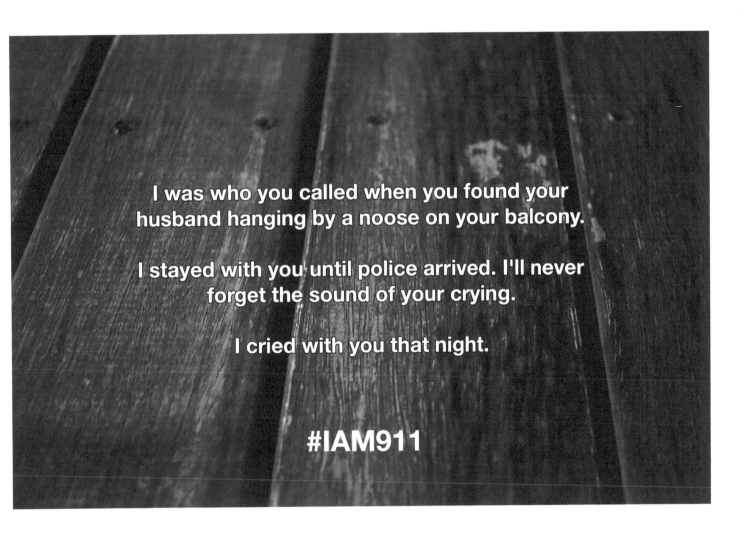

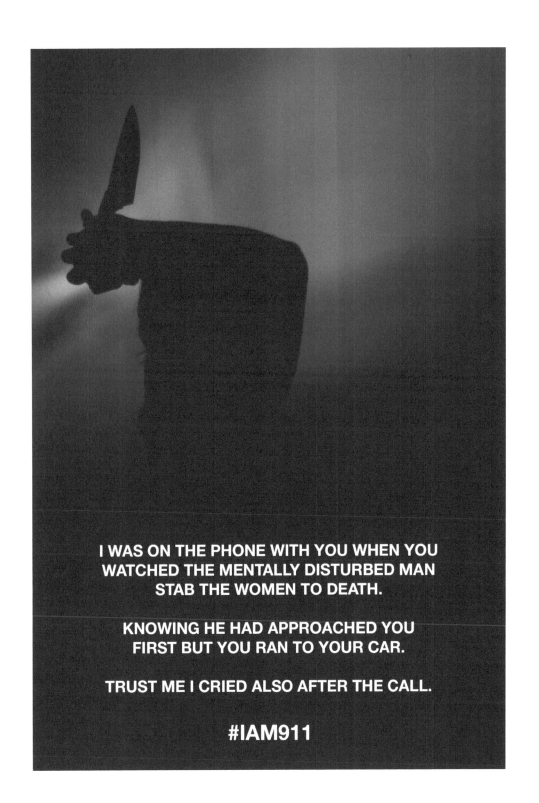

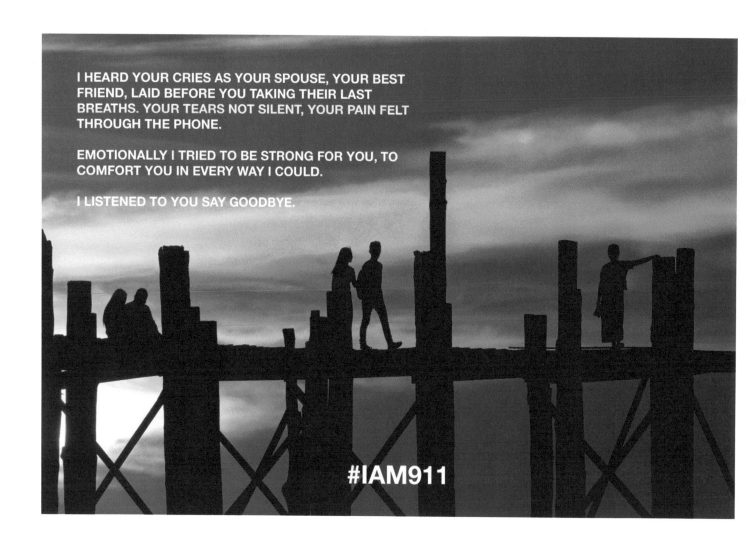

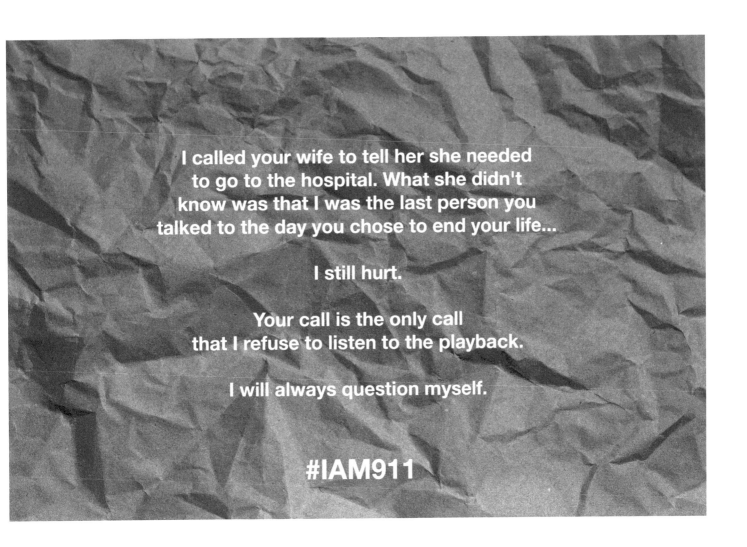

YOU RECOGNIZED MY VOICE AS SOON AS I ANSWERED YOUR CALL.

WE WORK TOGETHER.

I PROTECT YOU WHEN YOU LEAVE YOUR FAMILY.

MY VOICE GAVE YOU COMFORT AS YOU COMPRESSED YOUR INFANTS CHEST.

THE FEAR IN YOUR USUALLY COMMANDING VOICE HAUNTS ME.

#IAM911

HER**⬤**ES

(#IAM911 — IN THE MOMENT OF CRISIS I AM THE VOICE WITHIN THE TRENCHES)

We couldn't save him...

Out of all of my years in dispatch, there are two calls that dealt with brothers that rocked me. One evening I took a call from a man who I will call, "John." John and his brother had just left the funeral of a family member. His brother wasn't feeling well and slumped over in the passenger seat.

John quickly pulled over and checked on his brother. He was unresponsive. When I received his call he was frantic, as anyone would be in this situation. I got his location and sent out an ambulance. I asked him if he could pull his brother out and onto the ground so that we could start CPR.

"No," John replied.

"John, in order for us to do CPR he needs to be lying flat on his back, on the ground."

"I CAN'T GET HIM ON THE GROUND! I CAN'T GET HIM OUT OF THE CAR. PLEASE GOD NO! NO GOD!"

These are the moments where you feel helpless being on the other side of the phone. Your adrenaline is pumping and you're at the edge of your seat. Help was already headed out but CPR needed to start.

"John, is there any other way to get him out?"

"NO! I can't get him out. Can we try CPR?"

All of this is happening within seconds. Quick decisions are being made although it feels like forever is passing by. From the moment John said that he couldn't get his brother out of the car and onto the ground lying flat for CPR, I was already thinking of every possible way. I tell him something. I know it might not work, but John needs something, anything, to feel like he is helping his brother.

"Grab the lever on the right side of the seat and push the seat back as far as it can go. We need to get him as flat on his back as possible."

I know this is a long shot but if it were me, I would want the person on the phone to try everything in the book to help.

"Tell me when you've done it."

I can hear John quickly pushing the seat back.

"Don't do this. Don't do THIS," John says to his brother.

"John. Did you push it back?"

"YES."

We start CPR. All of this happens in seconds. At the time, as I've mentioned already, it feels like hours are going by. I can't see what is going on with my eyes, but the stage has been set in my mind. Everything I hear I am imagining. It is playing out in my head. I have put myself in my callers' shoes and the thought of my own brother emerges. Please let John's brother live. Please…

I'm sweating as we go through the process of CPR. Again, I don't know if this is going to work but I had to try something. The ambulance is almost there. I can feel the pain that John is going through. In between rescue breaths and compressions, John tells me that they had just left the funeral of a family member and he can't lose his brother too. I'm crushed on the inside. The ambulance arrives but John's brother doesn't make it. After the call I sat there, mourning the loss of John's brother as if he were my brother. Later on I was told that there was nothing I could've done. No amount of CPR would've saved John's brother. Although I know there was nothing more I could do, I still beat myself up over it. It's something we all do in this profession at one time or another.

John, I'm so sorry we couldn't save your brother. Wherever you are at, all these years later, I hope you are at peace. You did everything you could do and I was there with you.

Imagine Listening

Your worst day is our everyday...

12 Stories

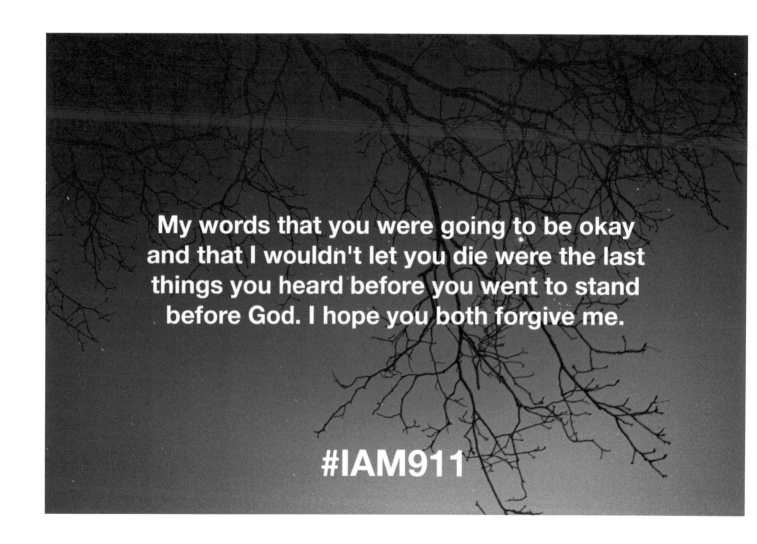

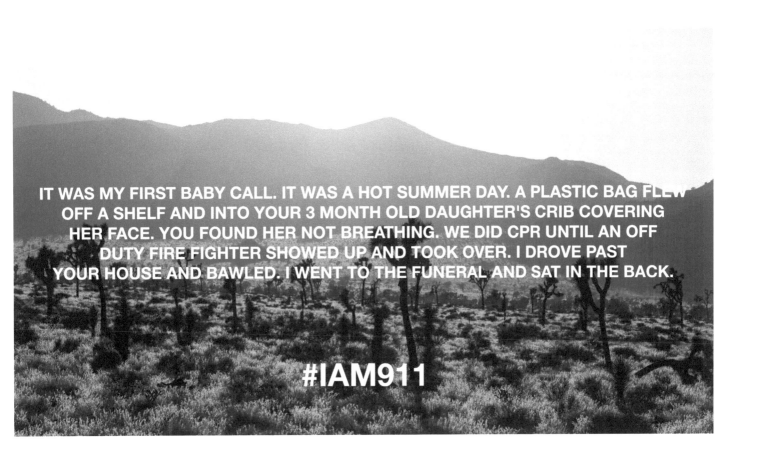

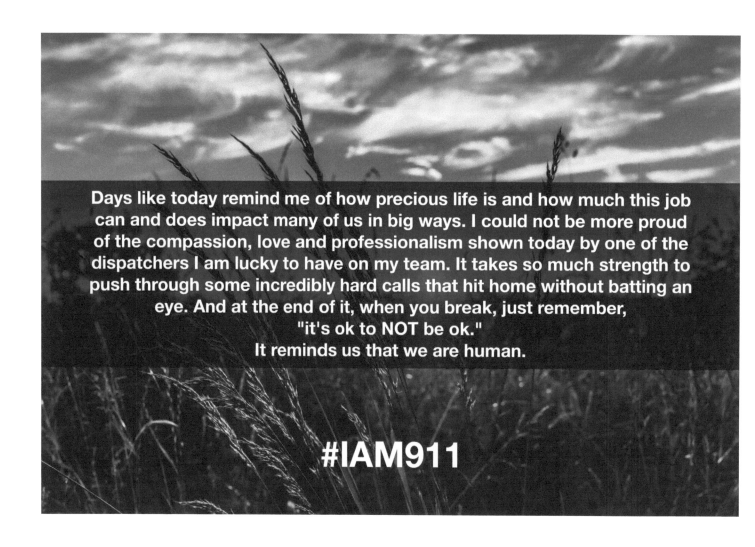

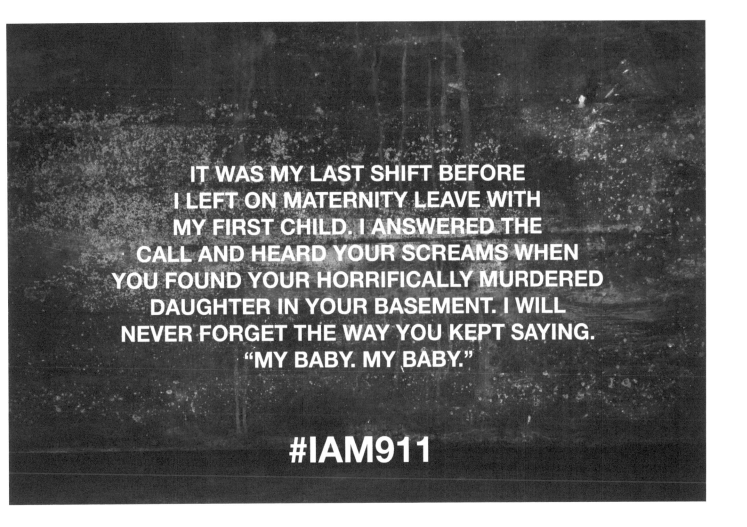

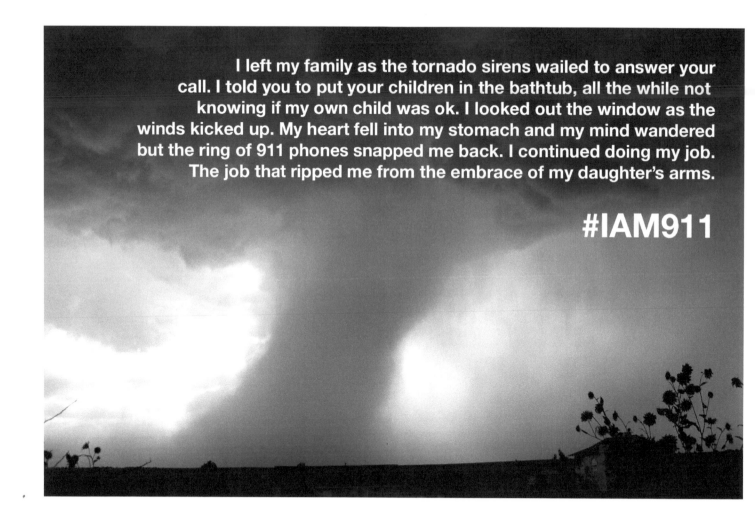

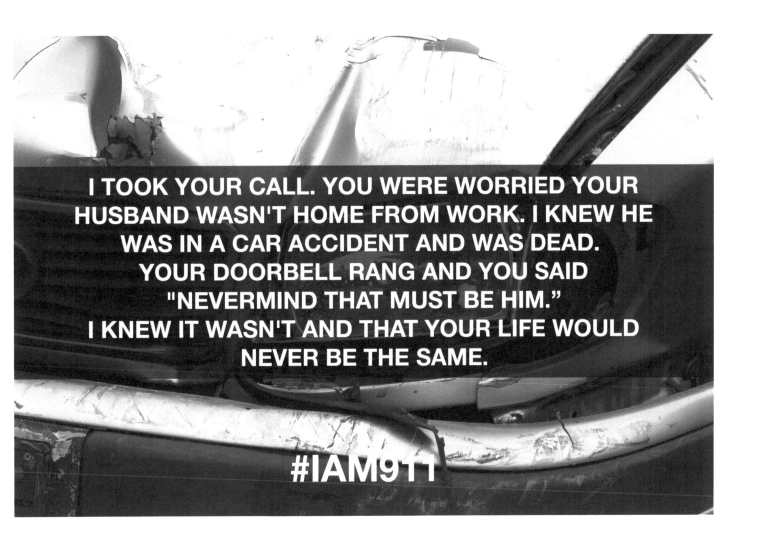

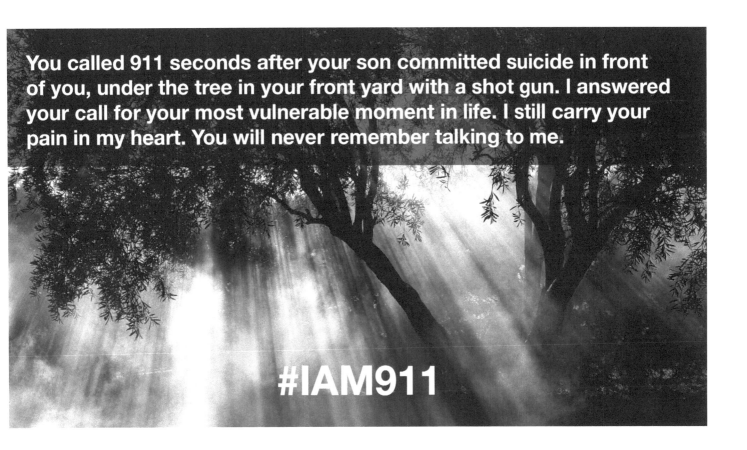

I was on the phone when your best friend was screaming at me to help.

I heard your last breath.

You cheered with my son.

Yesterday was your anniversary and I couldn't bear to see your family's pain.

This will never leave my heart and mind.

I'm sorry I couldn't fix this.

#IAM911

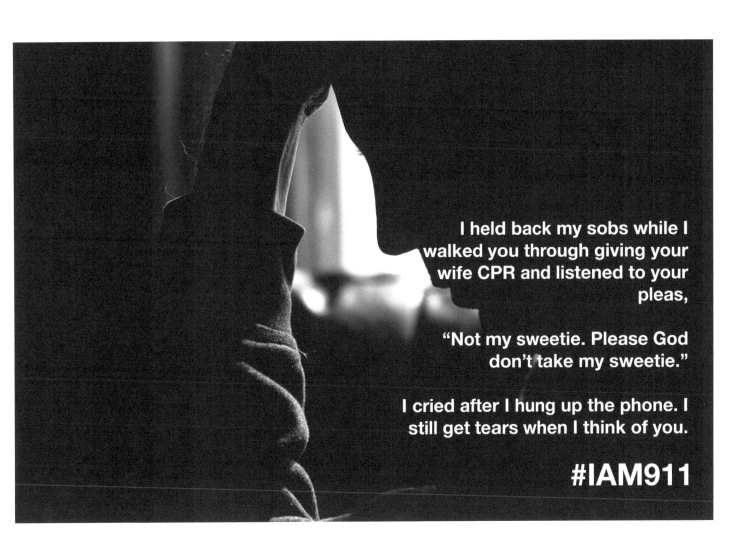

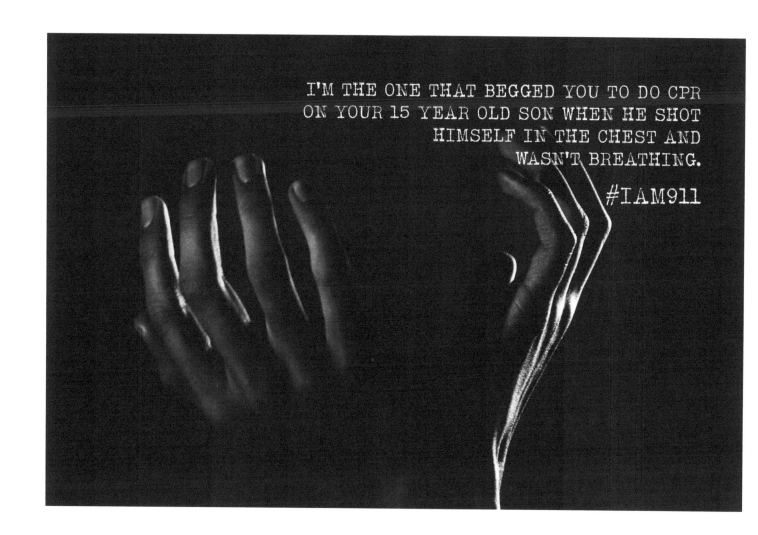

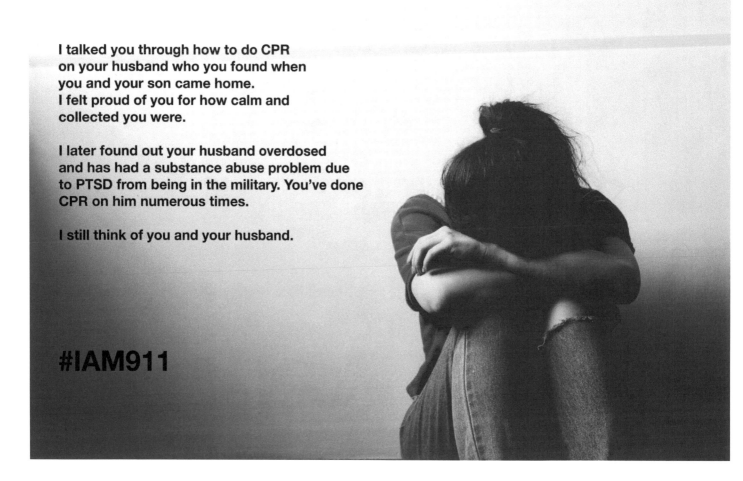

I talked you through how to do CPR on your husband who you found when you and your son came home.
I felt proud of you for how calm and collected you were.

I later found out your husband overdosed and has had a substance abuse problem due to PTSD from being in the military. You've done CPR on him numerous times.

I still think of you and your husband.

#IAM911

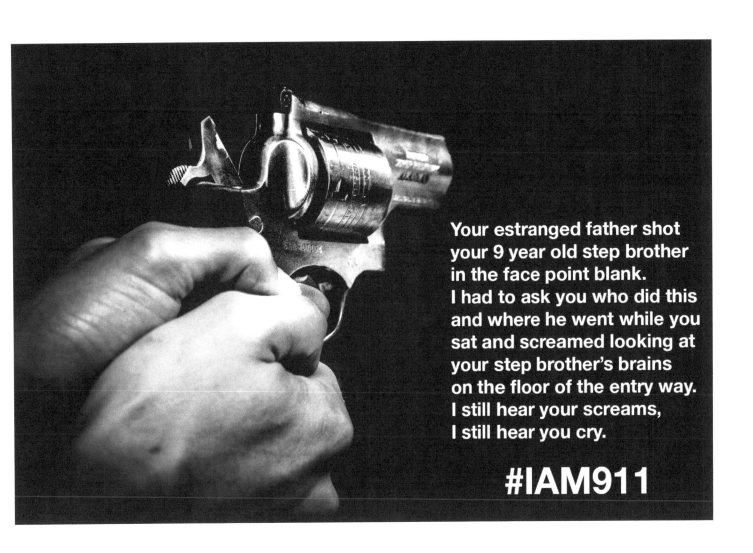

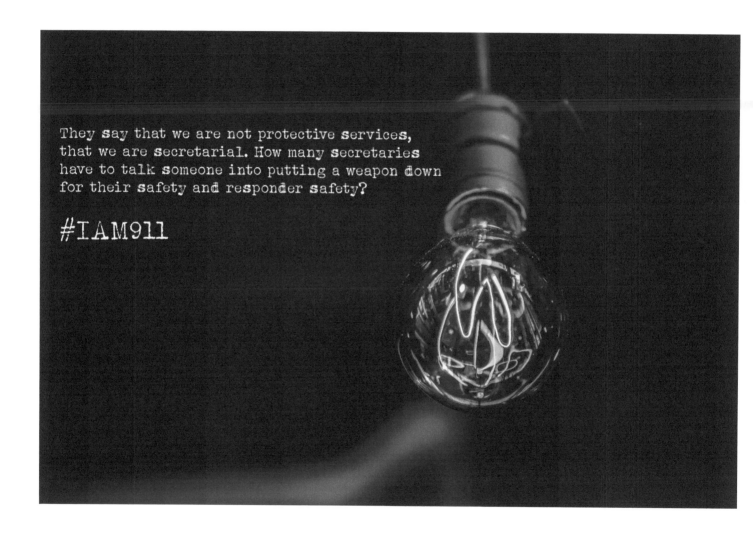

I received your call for shots fired, people down. You were screaming and pleading for your daughter to fight and keep living. The shooter was your daughters fiancé and he had killed your husband and injured her mother in law and your daughter. At the time, you didn't know you were shot as well.

In your arms you had your 10 month old granddaughter. The shooter was still there coming in and out of the house. I was on the phone for 15 minutes feeling so helpless and so scared for all of you.

A piece of my heart broke that day and I was never the same after that. Your daughter, who was only 25, and your husband didn't make it.

You, the mother in law, and your granddaughter survived.

PTSD is real for dispatchers. Don't ever feel ashamed to ask for help. I almost lost my life.

#IAM911

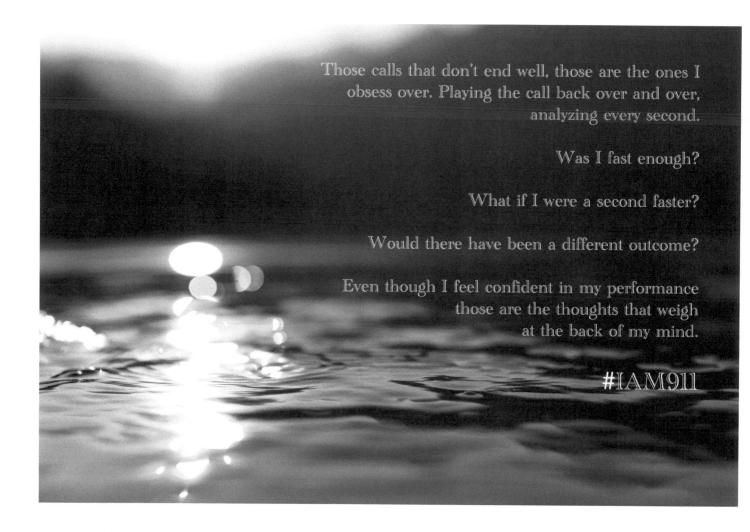

They say you can hear a smile. In my job I love dealing with challenging callers. Generally, I am good at it and can get an angry person to end up thanking me as we are disconnecting. However, the public can tell when you are having a bad day. They may not recognize that's what it is but they notice when your tone is off and sometimes it seems personal to the caller. I know they don't realize that it is actually me having only slept two hours. Me being awake to handle unexpected events during the day but still having to work night shift. Me crying as I come into work because I am overwhelmed. They don't know that I still show up to handle their emergencies while my family is dealing with its own crisis. Tomorrow is the beginning of National Public Safety Telecommunicators week so here's to us for always being there.

#IAM911

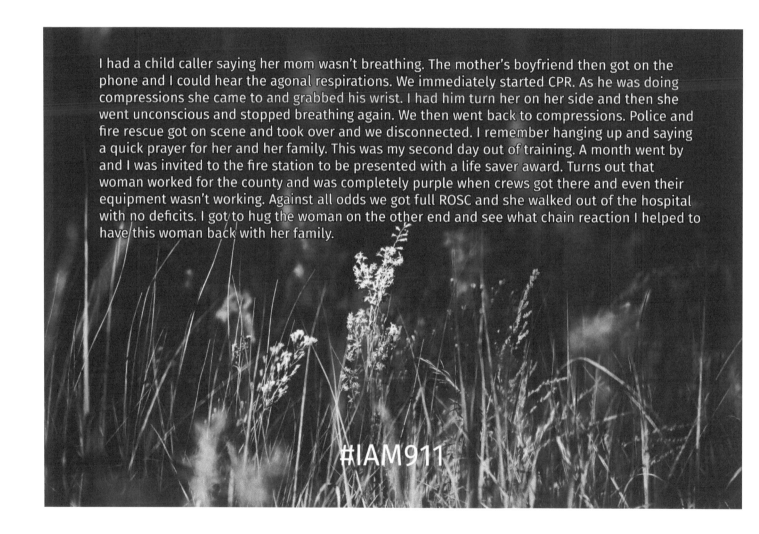

I had a child caller saying her mom wasn't breathing. The mother's boyfriend then got on the phone and I could hear the agonal respirations. We immediately started CPR. As he was doing compressions she came to and grabbed his wrist. I had him turn her on her side and then she went unconscious and stopped breathing again. We then went back to compressions. Police and fire rescue got on scene and took over and we disconnected. I remember hanging up and saying a quick prayer for her and her family. This was my second day out of training. A month went by and I was invited to the fire station to be presented with a life saver award. Turns out that woman worked for the county and was completely purple when crews got there and even their equipment wasn't working. Against all odds we got full ROSC and she walked out of the hospital with no deficits. I got to hug the woman on the other end and see what chain reaction I helped to have this woman back with her family.

#IAM911

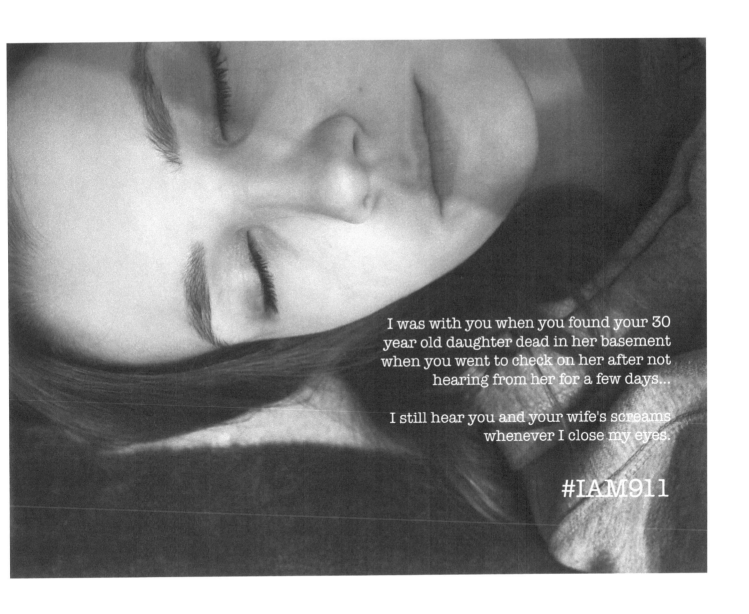

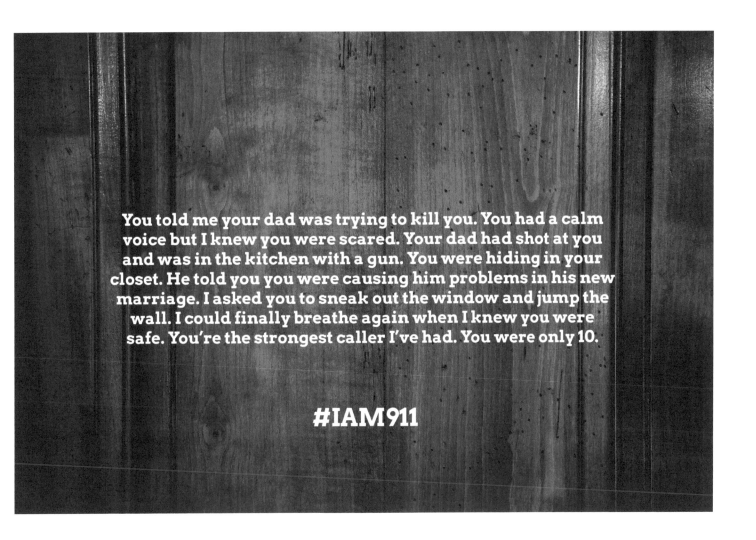

When you first called in to 911, you had said your husband was shot. You had walked in the garage and seen him sitting there with blood coming from his mouth.

I remember you called from the landline and I got the police going. I heard you wailing and screaming that he had the shotgun in between his legs, and how were you going to tell your grown children what had happened.

I stayed on the line with you, and tried to help you to breathe in between the screams. I have never looked at an entrance to a garage the same since that day, and the images I have in my head haunt me each day.

My thoughts are with you and your kids, the generational trauma sticks with me.

#IAM911

SHOTS FIRED

In 911 we hear the worst of the worst. People are not calling us because they are having a great day. I don't ever remember anyone calling to say,

"Hey! I just wanted to call and tell you to have a great one."

It doesn't happen. People are calling while they are living their worst nightmare. Occasionally we get a thank you. We are recognized for what we do, and it's not that we need all the kudos in the world, but a thank you now and then from our callers or those out in the field is always welcome. Although there are good moments in 911, the majority of what we deal with is horrifying and scary.

One of the scariest things to hear doesn't come from our callers. It comes from our officers. It's something that no one ever wants to hear over the radio. One night, I was on main law enforcement radios. One of my state troopers checked out at a residence on a welfare check for a female in her early 20's who was with her boyfriend. Her parents were concerned, called the post and requested a welfare check on her because they could not reach her.

When the trooper arrived at the home, he explained what he was doing and said, "I'll be 10-5." Now, ten codes are different for almost every agency, and when it comes to the state, at least where I was, they used the most common ten codes. 10-5 meant that he didn't need any status checks, but for us it also meant, set a timer for 15 minutes and then check status. My trooper was on scene for less than five minutes when he yelled over the radio,

"SHOTS FIRED! SHOTS FIRED!"

My body went cold. I sat at the edge of my seat, pressed the button on my radio and repeated the information, gave the location, and called out, "Central will be under Code 7, all other traffic will move to secondary."

As I type this, I am sent back to that moment. My mouth went dry. My heart was pumping and I prayed that my trooper was ok. My team and I went into full gear, working like a well oiled machine. Officers in the area said that they were headed to the trooper's location. When this happens you are on autopilot. Every detail of what is going on is typed into your computer for everyone to see on their laptops and in the emergency communications center. As I type and relay information over the radio I hear one of my partners say,

"Oh my god you guys."

We continue to work. This was the second time I had heard one of my officers yell out shots fired. My emotions were in the back of my head. I can't break because my team and my officers need me to be strong. It's one of the hardest things to hear and deal with but you push through. You get it done. This had all happened just before our shift was over. I remember thinking,

"I wonder why the family waited until very early in the morning for this welfare check?"

I figured that they felt that someone would be home rather than during the day when they might be out and about and a welfare check would be pointless because of the possibility of no one being there. While all of this is going on we find out that our trooper had been shot in the hand. He was ok. When the trooper made his way to the bedroom to check on the daughter of the parents who had called the post, the boyfriend shot at my trooper. Right before it happened, my trooper could see the girl slumped over the bed. In that moment of chaos he couldn't say if she was ok or not. He had been checking the house, following protocol, and made his way to the bedroom but when the door opened… "SHOTS FIRED! SHOTS FIRED!"

Backup arrived to assist my trooper. A perimeter was set up and the unthinkable happened.

"You guys… Do you think the suspect will call in?"

There is always a chance that the suspect will call in.

I have worked numerous times when a suspect or person of interest has called in. It can be a bone-chilling experience. In 911, it seems like if you speak it out into the universe, the universe will respond. No sooner did my partner say this, the suspect calls saying that he could see our officers in the driveway. We let our officers know that the suspect was on the line with us. We tried to talk to him but he ultimately hung up.

By this time we were also working a fatal car accident. All of this was happening at the same time because 911 doesn't stop. Just because we have one major emergency going on doesn't mean that another one won't happen. Anything can happen in 911. Our relief had arrived but we didn't want to leave. This was our incident but after some convincing, we did leave. I remember driving home in silence, praying that the girl in the room was ok and that my officers would be ok as well.

When I got home I had to force myself to sleep. So many things happened and it still wasn't over. I needed to know the outcome, but I also needed to rest. When I closed my eyes they opened right back up. I was exhausted. I had a few hours of sleep but it felt like I blinked. I ran over to my computer and searched the news for an update and there it was. Pictures had been posted from the news crew that was at the scene and my heart started pumping again. SWAT had been called out for the standoff and I read the article while holding my breath.

"Please let the trooper and the girl be ok. PLEASE."

The trooper was going to be ok. He was indeed shot in the hand, but he would recover physically. The suspect was taken into custody, but the girl did not make it. According to reports, it is believed that she may have died before officers arrived. To this day, it is a call that sticks with me. It was the second time in my career that I had heard one of my officers yell over the radio. It's one that is hard to forget, but I am grateful for my team and my brothers and sisters in the field. We had started discussing our calls more and more around that time and it helped us get through it.

The suspect was later sentenced to life in prison for the murder of his girlfriend.

Imagine Listening

Your worst day

is our everyday...

Vol. 2

Stories

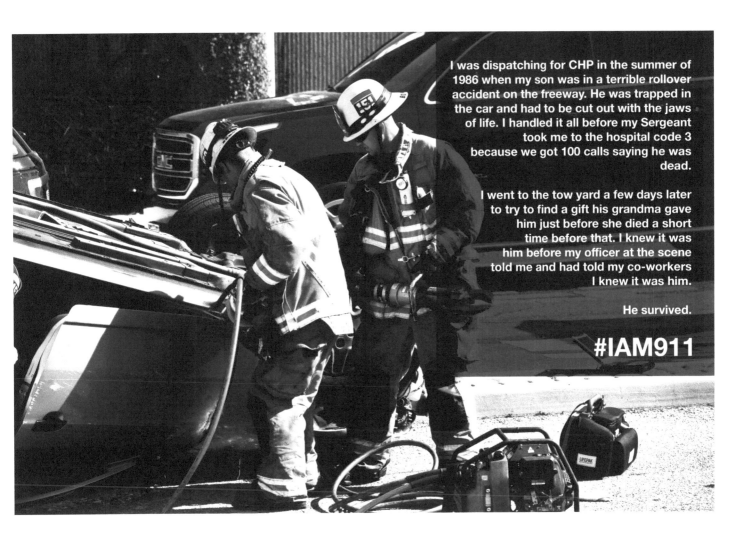

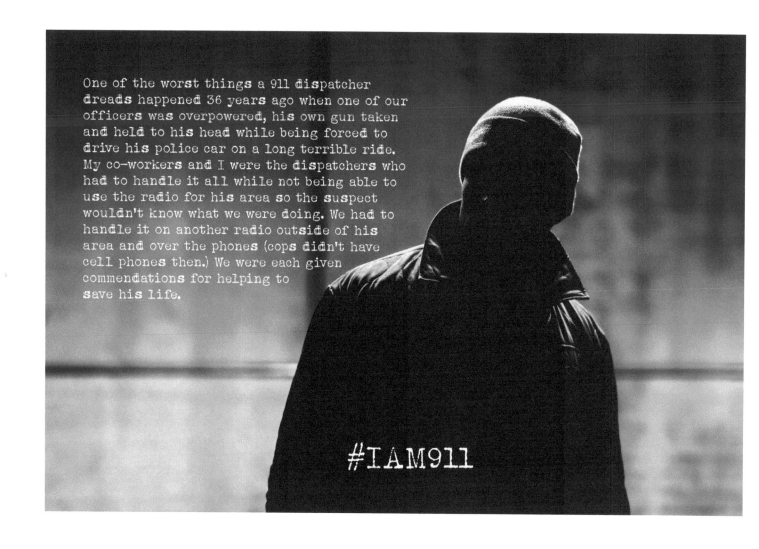

One of the worst things a 911 dispatcher dreads happened 36 years ago when one of our officers was overpowered, his own gun taken and held to his head while being forced to drive his police car on a long terrible ride. My co-workers and I were the dispatchers who had to handle it all while not being able to use the radio for his area so the suspect wouldn't know what we were doing. We had to handle it on another radio outside of his area and over the phones (cops didn't have cell phones then.) We were each given commendations for helping to save his life.

#IAM911

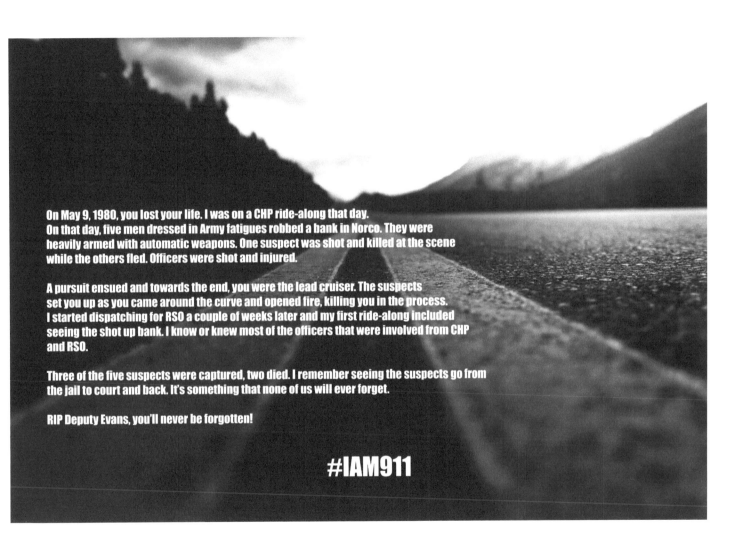

On May 9, 1980, you lost your life. I was on a CHP ride-along that day. On that day, five men dressed in Army fatigues robbed a bank in Norco. They were heavily armed with automatic weapons. One suspect was shot and killed at the scene while the others fled. Officers were shot and injured.

A pursuit ensued and towards the end, you were the lead cruiser. The suspects set you up as you came around the curve and opened fire, killing you in the process. I started dispatching for RSO a couple of weeks later and my first ride-along included seeing the shot up bank. I know or knew most of the officers that were involved from CHP and RSO.

Three of the five suspects were captured, two died. I remember seeing the suspects go from the jail to court and back. It's something that none of us will ever forget.

RIP Deputy Evans, you'll never be forgotten!

#IAM911

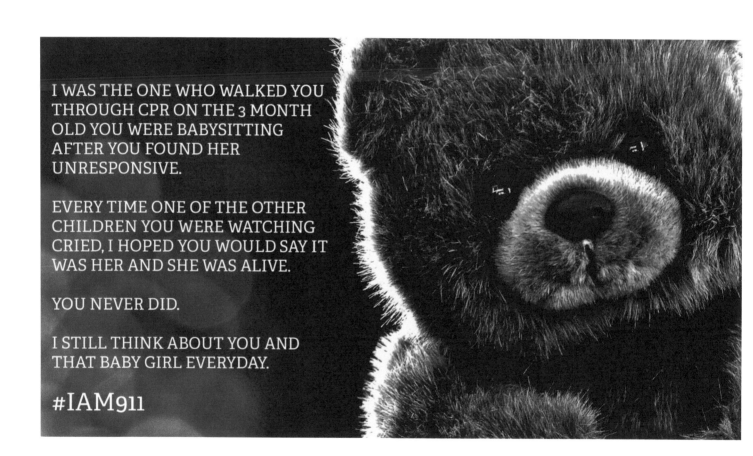

I WAS THE ONE WHO WALKED YOU THROUGH CPR ON THE 3 MONTH OLD YOU WERE BABYSITTING AFTER YOU FOUND HER UNRESPONSIVE.

EVERY TIME ONE OF THE OTHER CHILDREN YOU WERE WATCHING CRIED, I HOPED YOU WOULD SAY IT WAS HER AND SHE WAS ALIVE.

YOU NEVER DID.

I STILL THINK ABOUT YOU AND THAT BABY GIRL EVERYDAY.

#IAM911

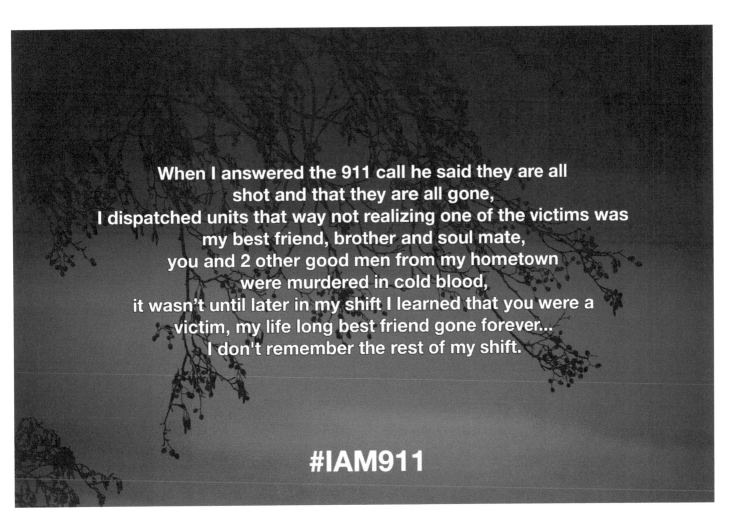

When I answered the 911 call he said they are all
shot and that they are all gone,
I dispatched units that way not realizing one of the victims was
my best friend, brother and soul mate,
you and 2 other good men from my hometown
were murdered in cold blood,
it wasn't until later in my shift I learned that you were a
victim, my life long best friend gone forever...
I don't remember the rest of my shift.

#IAM911

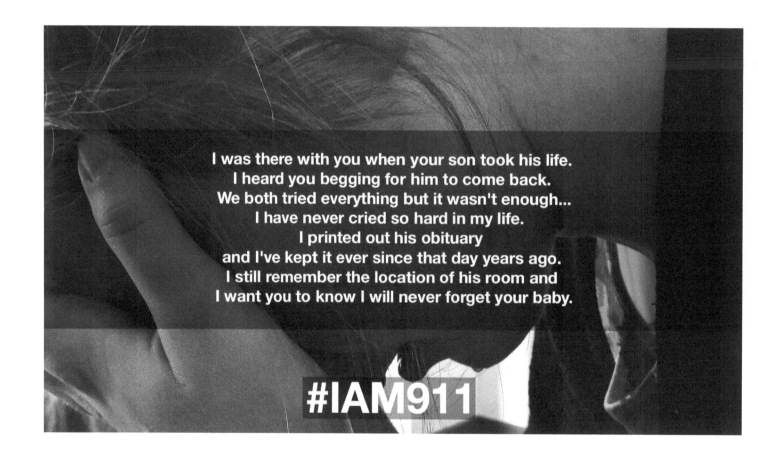

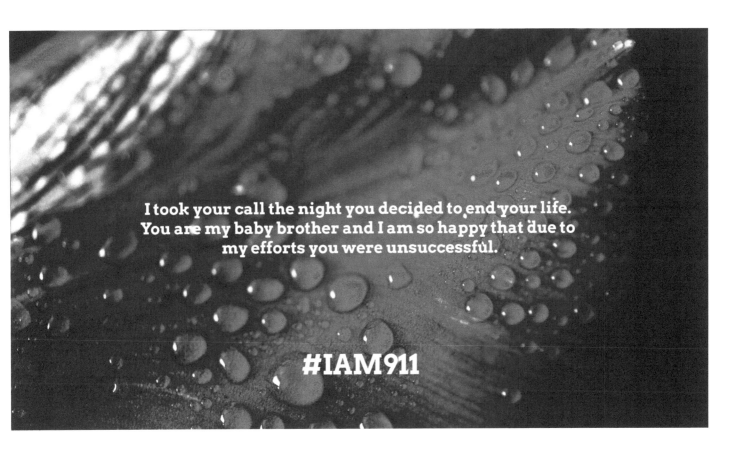

I was the last person to hear your voice. I was the last person's voice you heard right before your husband of many years shot you. You were so confused but yet so calm when you told me he had a gun. I heard the single gunshot, I heard his cold voice tell you we would never make it in time. I already had officers rushing to you but they were too late. Before they could get to your body your husband shot himself as he lead them to you. My next call was your son. I hear your voices everyday.

#IAM911

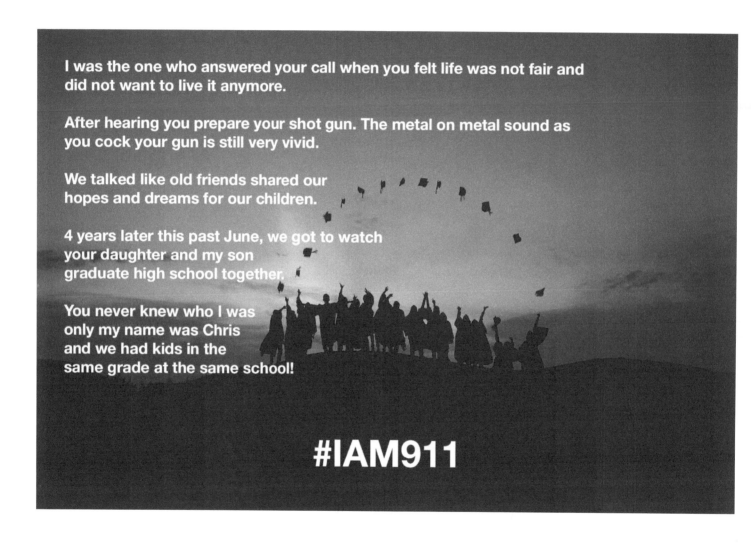

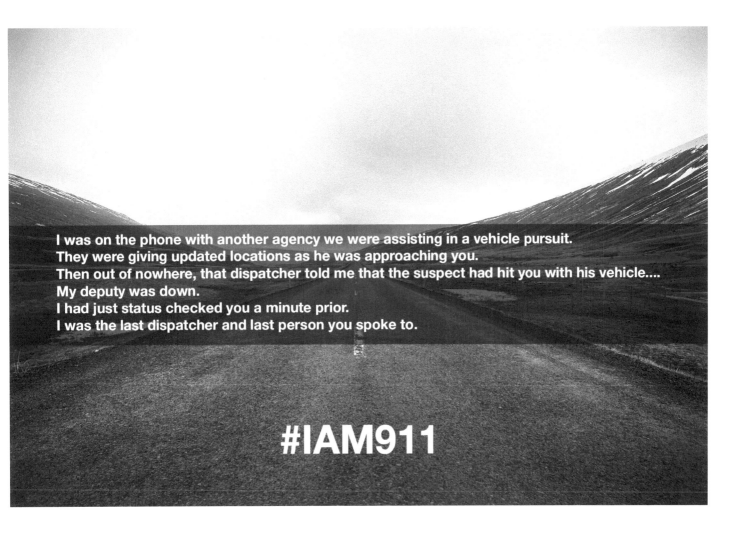

I spoke to you the night that the Veteran's Hotline called because you were suicidal. You kept disconnecting with them. Their hope was that we could locate you. I tried calling you and you answered. I told you I wanted to help and you told me I would never understand. I said that you were right. I would never understand, but that I would forever be grateful for the sacrifices that men and women like you make for people like me.

We spoke for over an hour and during the conversation I was able to piece together where you were and what jurisdication. My partner had that agency send a military veteran to check on you. I waited on the line with you until that officer arrived.

I'm sorry I didn't tell you right away what I was doing, but I knew you had your gun and you were ready to go, and I wasn't going to give up; I cared too much.

You began to cry and I cried with you. I was able to convince you to go outside, unarmed, to meet with the officer. The last thing I heard was you asking the officer for a hug and him responding with, "yeah."

I think about you often and hope that you are still out there safe. I pray that you never again have to go through the battle you went through that night.

I'll never know if you did, but I'll always know that night I helped you win that battle.

#IAM911

We discussed the challenges of the career together. We had a group to support each other. You were brave enough to leave all your support and strike out into a new state 1000 miles away to try and create a fresh start. While you were there you became sick and lost your hearing which led to you not being able to dispatch anymore. By the time this happened it was woven tightly into who you were. In the background you were being physically, emotionally and psychologically tortured by someone you thought you could trust. Control issues, animals you cared about stripped from you and your home taken away. You fought like hell to get back on your feet and try to track down all that was taken AND you did. That fight was brutal, you could hear it in your voice when we spoke to each other. You settled into a job at Target in the fraud department hoping it would be challenging enough. In your soul you knew it wouldn't be.

You decided to consult but at times it just reminded you of the career you could no longer have. When conferences would come you would go off grid because it hurt too much to be reminded of your loss and the absence of time around the network that loved and supported you. I understood but always worried. To know you was to immediately love you. You were larger than this world could contain. The fighting to hold on just became too much and no one could reach you. When you took your life, my world was rocked as well as those around me. I cried inconsolably throughout the night and the first week and anytime I see something that reminds me of you over the last 9 months. There is no day that goes by that I don't see a sunset, some landscape or water feature that I know you could capture via lens.

I miss you, my friend.

Because of you I train differently and I protect them fiercely. I cannot express enough how outlets, support networks and the separation of personal identity and 911 identity is crucial to those in this field. I hope you know; you have changed us. I just wish you were here to see it. You are not my first colleague who lost their life by their own hand, but it has to stop.

#IAM911

I took your call for your 6 week old granddaughter not breathing. I begged and pleaded with you to keep trying CPR until the volunteer department or my deputies could get on scene to help. I did everything I was trained to do, and more, to encourage you to keep doing compressions. And when you stopped too soon, I consoled you through my own frustrations.

Afterwards I sobbed in my car my entire lunch break.

No one knows, but I've been struggling with fertility for a long time. My husband and I have tried nearly everything without success. I felt your loss as deeply as my own and I still hear your cries months later.

#IAM911

The dispatcher I called, after six years of running or being abused, was the only person who really did anything to help me escape.

Now my daughter and I are finally safe from my husband, and her father.

I wish I could thank her.

#IAM911

You called in the middle of the night and told me you were going to kill yourself.

You said you had no reason to live and no one cared.

I was there for you and talked you out of it.

I'm glad you changed your mind.

#IAM911

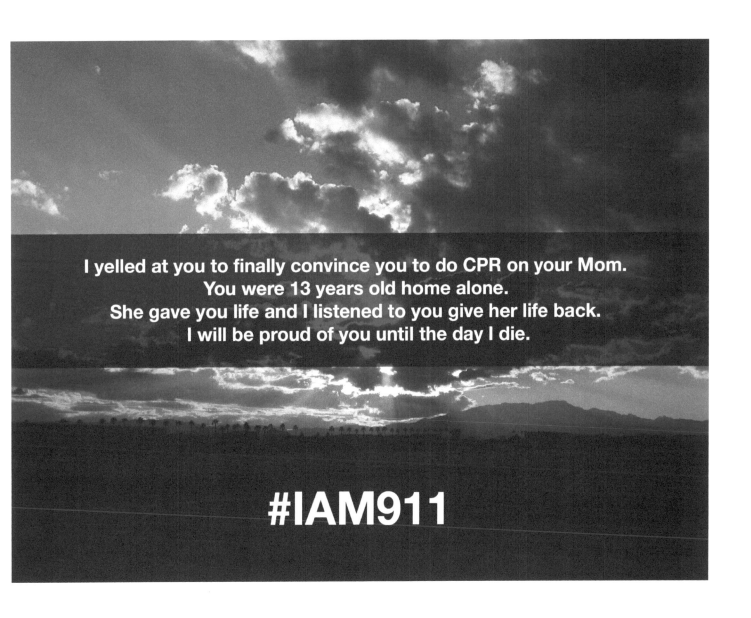

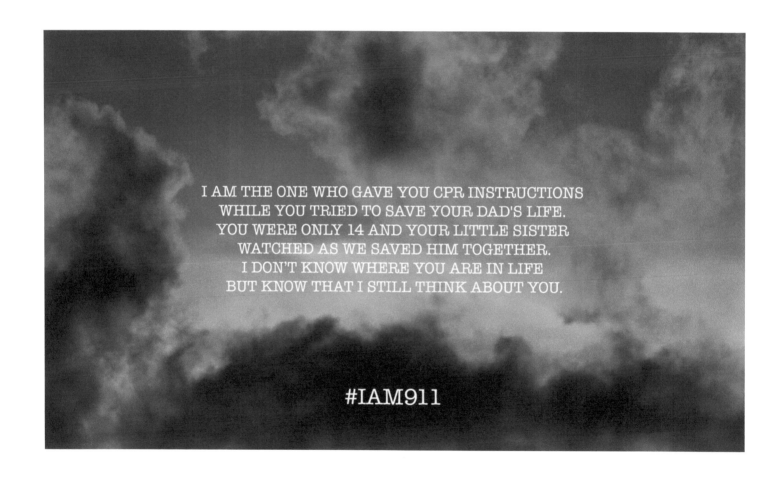

#IAM911

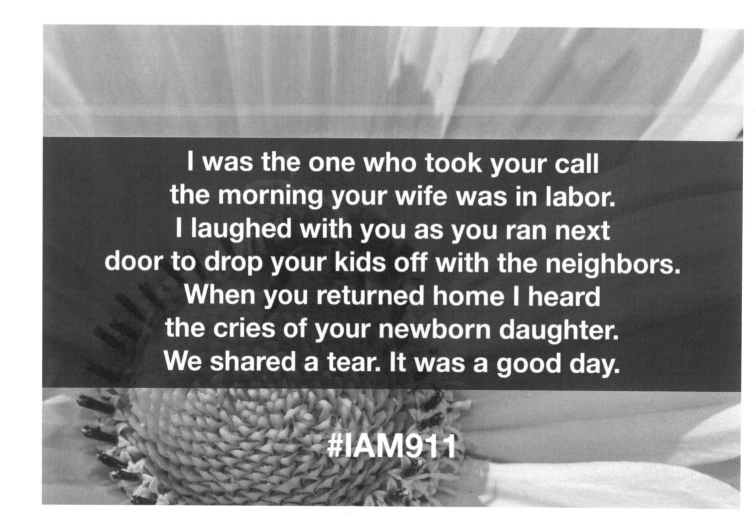

Sir, where are you going?

For all of the horrific things that we hear in the profession of 911, there are also some sweet, and rewarding moments. They are sometimes few and far between, but when it happens, it is the most amazing experience. One such rewarding moment is when you receive a call for a woman in labor.

Now don't get me wrong. This type of call is nerve wrecking. You can have all the training in the world, but when this call comes in, there is no amount of training to prepare you for it. What it takes is doing it over and over. In some emergency communications centers, that doesn't always happen. Some people can go through the majority of their career before taking a call like this.

I have taken many calls like this, however, one morning, I received a call that will always stick with me. Years ago, on the fourth of July, I had a new trainee with me. We were talking about her work experience and I was showing her some of the equipment we use to do our job. As I was telling her a little bit about my background, a 911 call came in.

"911, where is your emergency?"

"I NEED AN AMBULANCE! MY WIFE IS IN LABOR!"

Right off the bat the adrenaline kicks in. I get the information needed to send an ambulance and my caller yells,

"THE BABY IS CROWNING!"

Everything is happening so fast. I already have an ambulance headed that way and I am going through our emergency medical dispatch protocol. My trainee's eyes are popping out of her head as she listens to what is going on. I'm trying my best to keep my caller calm while keeping my excitement at a reasonable level. Up until this point I had received several calls like this, but none have come close to where the baby has been born, which would mean that I would receive a stork pin for helping deliver a baby over the phone. This was my shot!

"Huh, huff, huff, huff."

I stop for a moment. I no longer hear my caller's wife in the background. Instead I hear huffing and puffing.

"Sir? SIR? What are you doing?"

"I'm sorry," my caller yells while out of breath.

"You're sorry? What do you mean," I asked.

"I took off," he responded.

"You left your wife alone?"

"I had to! I needed to drop my kids off at the neighbors."

I understood. Next I heard him running back to his place to be with his wife. EMS was just pulling up when I heard my caller run into the home and the baby was born. We chuckled a little at the situation and I began to tear up. I was so excited for them. My caller quickly hung up as I congratulated them. Because my caller left I didn't receive the stork pin. I was bummed but it was such a memorable experience that the pin was secondary. I was happy that I could be there for them. I later found out that it was a little girl. I wish I had known while on the phone with them so that I could suggest the name of "Ricarda."

That call will always stick with me because of the entire experience. It was a roller coaster of emotions, but I wouldn't change it for the world. As for the stork pin, I never took another call that was that close. However, years later, I would share this story, LIVE, on social media, at a dispatch center in North Carolina. As I finished up my story, someone with the agency walked into the room where I was and handed me a stork pin. I began to tear up and I was emotional.

"This is one of our stork pins. We were listening to you tell your story and we wanted to give you one of ours because you should have received it back then."

Thank you Orange County. You have no idea how much that meant to me.

Orange County, NC – Stork Pin

If I could

tell you one thing

If I could tell you one thing

Back in 2019 I had a conversation with my wife about 911, family, and friends. Prior to this conversation we had recorded an episode of the Within the Trenches podcast where we discussed what it was like for her and our children while I was working in dispatch.

One of the things she asked was what I might have wanted her to know while I was struggling with my hard calls, and my burnout stage. As we continued to talk she mentioned that others might want to know the same thing. I thought it was an outstanding idea.

I wasn't sure if anyone would contribute, but if it was done anonymously, maybe others would share. I asked 911 professionals what one thing they would tell their friends or family about their career. It resulted in #IfICouldTellYouOneThing. Thank you, Rebecca, for the conversation, and inspiration.

Now that you have read countless stories from those within the trenches of 911, read what they would like you to know. The following thoughts and feelings set the stage for you, a family member, friend, co-worker, husband, wife, reader, etc. to experience a deeper connection with each 911 professional.

When you take this job, there's a line you cross that you can
never cross back over to be normal again.
And your friends and family don't understand/get it.
It's like being on the other side of a wall and not fully being able to be 100%
involved/present in anything else in your life because you are reserved now.

I WANT PEOPLE TO KNOW THAT WE ARE NOT GLORIFIED SECRETARIES. WE ARE THE FIRST FIRST RESPONDERS. WE SIT IN SMALL SPACES AND USUALLY DON'T HAVE A LOT OF TIME TO STOP DURING OUR SHIFTS. A LOT OF TIMES WE DON'T GET TO SEE THE END RESULTS OF OUR CALLS. THE NOT KNOWING IS HARD TO HANDLE.

I'd like to talk about the bad calls, but no one cares.
Everyone jokes about everything because that's our defense
mechanism, and the people that aren't on the job
don't want to hear about them.

I've been retired for a few years now. I like staying home. I don't like large crowds.
I don't know what to say to people.
I can't do small talk well.

SO MANY TIMES I'VE BEEN TOLD TO LEAVE WORK AT WORK, BUT I WILL ALWAYS SEE THE YOUNG CHILD AT THE BOTTOM OF THE POOL AND HEAR HER MOTHER'S SCREAM BEFORE THE HAUNTING SPLASH.

WHEN I COME HOME FROM WORK AND DON'T WANT TO TALK,
I'M NOT IN A BAD MOOD. I'M JUST TIRED OF
LISTENING AND TALKING AND
I NEED A BREAK.

I LIKE THIS. IT'S NICE TO NOT FEEL ALONE OR WEIRD FOR FEELING THESE THINGS. IF I COULD TELL MY FAMILY ONE THING, TO MY LITTLE KIDS WHO HAVE NO WAY TO UNDERSTAND... MAMA IS SORRY THAT SHE STRUGGLES TO LISTEN TO THE LONG STORY AND SHE DOESN'T HAVE ABUNDANT PATIENCE FOR YOUR PROBLEMS. MAMA STRUGGLES TO NOT JUST "SOLVE" AND MOVE ONTO THE NEXT THING IN OUR DAY.

I DON'T MEAN TO DISMISS WHAT'S IMPORTANT IN YOUR LITTLE WORLD. MAMA IS SORRY THAT LONG 12 HOUR SHIFTS SOAK UP EVERY LAST POSSIBLE DROP OF PATIENCE, EMPATHY, KINDNESS THAT I HAVE UNTIL I REST THE NEXT DAY.

I HATE IT THAT I CAN'T GET BACK THE BEDTIMES WHEN I HAVE NO GRACE BECAUSE OTHERS NEEDED ME ALLLLLL DAY AND IT LEFT MY CUP EMPTY. SOMEDAY I HOPE YOU GET IT, EVEN IF JUST ENOUGH
TO STILL LOVE YOUR MAMA.

MY DESIRE TO KEEP YOU SAFE LED ME TO A PATH OF SERVICE; FOR YOU AND EVERYONE WE LOVE.

DON'T BE AFRAID OF THERAPY.

WHEN I TAKE MY HARD EARNED VACATION, ALL I WANT TO DO IS RELAX AND DECOMPRESS. PLEASE UNDERSTAND THAT MY ONLY PLAN IS TO HAVE NO PLAN BECAUSE I JUST NEED QUIET TIME AWAY FROM THE CHAOS OF THE WORLD.

I'm sorry my anxiety and irrational fears affect your life so much and you don't understand it. This is why I don't want any of you to work in Public Safety in any capacity. I don't want you to be afraid of everything like I am. I want you to live normal lives, be happy, and be able to have relationships with people.

I am not trying to be pushy, overprotective or overbearing. I care about you and never want you to call 911 in terror or pain. My experiences on the other end makes me highly protective of those I love.
Please be patient with me and my attempts to keep you safe.

Anywhere there are unknown people, anxiety happens. Restaurants, stores, anywhere public we go. I automatically start making plans in my head on what to do if something bad suddenly happens. I find exits, I won't sit with my back to a door, and my head is on a constant swivel. I can't turn it off.

Don't mistake my, "not listening," for intentional disregard for your feelings. I just left a place where I listened to life's worst problems for 48 hours straight. Sometimes I just want to stare ahead and forget about it. Be patient with me.

Significant other relationships are hard and don't last long. No one on the outside knows or can understand what we go through even when we explain everything to them.

If I'm talking about what happened, and you're in a headspace to listen, then just listen. I'm not looking for answers or suggestions unless I tell you that I am.
Sometimes I just need to process what happened, in a safe place with you, the person I love and trust enough to tell.

It's going to hurt. The ones you lose and even some of the ones you save. It can make life difficult, including relationships and even just watching certain movies. But push through it, and talk about it. Don't keep it inside.
You got this.

It's not that I don't want to share my stories, it's that I don't want to put that image into your head. I live with it, so you do not have to.

I'm sorry I'm always on high alert. It's been embedded in me to keep my head on a swivel. I'm always prepared for any event to take a turn for the worst.

I'm sorry that I don't react as you want me to for certain situations. I've been trained to remain calm and handle things with a clear mind.
It's not due to a lack of emotion.

Everyone wonders why I can't yell at my children when they're bad… but that child in the car seat who was left in a cold car all night & he's no longer with us sticks in my mind every day...

No one teaches you in school how to deal with
the traumatic events you encounter.
We may come home agitated or quiet. It's not you.
We just can't let our emotions out when we need to, we
instead keep a prioritized mindset and do our job,
and move on to the next call.

We keep our eyes and ears open in public with exit routes planned.
We aren't paranoid, we just have to keep ourselves and our families
safe and prepared, whether our families
realize we do it or not.

We don't expect you to know what we actually encounter, and honestly we don't want you to. We don't want to expose you to the harsh reality of our careers. We do however ask that you just listen to us while we talk/vent, and be supportive of us.

When joining the first responder field you might lose a friend, or two due to the schedule or due to how the job will change you personally, but the friendships you make are so worth it. It's like having a whole other family you go to work with.

The sleep medication doesn't help. The talking about it doesn't help. I still hear the sounds of that screaming mother seeing her child move into the next life. The sound of her soul being ripped from her chest.

I am sorry I don't have much emotion to things. I have been trained to not show emotion. It's not that I don't care, I just don't show it how a normal person would.

Even though I left less than a year ago the mindset follows me still. Get to the pertinent information quickly. I don't mean to be short and sound uncaring, but it's hard for me to turn it off. I'm not good at conversing anymore.
It just feels like all I am capable of is small talk.

When people ask about how to describe my job as a 911 telecommunicator and how mentally we handle it, I always answer with this. "We hardly even know the outcome of calls, and it slowly chips away at you. It's like reading a horror story and when you get to the last chapter, you close the book and never find out how it ends. You always then worry for the caller and their family and you take that worst case scenario home with you every night."

It is physically and mentally impossible for me to turn work off in my head when I'm not there. Every day I relive situations in my head, thinking about different outcomes and things that I could have done different.

I give my all to everyone but sometimes things don't work out the way you want and that has been the hardest thing I've had to come to terms with. Some days are better than others but all I ask is that you're patient with me.

When you join the first responder world, nobody can prepare you for the fact that you are quite literally the difference between someone living or dying. Everyone looks to you to know what to do. And even if you don't know, you can't show it. It's an insane amount of pressure on a person.

As first responders, we are still human. We are too often told that we need to have thicker skin, and while that is true, the things we see on a daily basis do catch up no matter how hard you try to "leave it at the door."
Please be kind and patient with us, we are trying our best.

I know I love hard. I know I tell you I love you at the end of every call or text. When I tell you I love you over and over it is not because I am needy or trying to be annoying. I tell you many times because of the voices
I have heard on 9-1-1 over the years.

Some from loved ones screaming for those they lost, others wanting to end it with me on the phone, and I think about how this person just wanted to hear a voice, to have someone there with them.

Did they get to hear, "I love you" before living their nightmare? I tell you over and over because it would haunt me for you to not hear it. Working in 9-1-1 shows you how short life can be. This is why I love so hard.

I love you, the most amount.

Please don't get upset with me when I would rather text than talk on the phone.
Sometimes texting is all I can do.
Answering 911 for years has made it hard to have long
conversations over the phone. Please be patient with me.

I know it was hard for you to understand when you were younger. I'm so sorry that I couldn't be at all of your school functions, family outings, and holidays. I know it was hard on you but every moment I had, I spent with you.

Sleep didn't matter as long as I had the chance to spend time with you and play games. Although it was hard then, it means the world to me that now that you are older, you understand everything. I felt like a horrible parent back then. When you told me that you understand now and that I am amazing for the work I did, it meant more to me than you will ever know.

Years in dispatch has trained me to get the important information fast.
I need specifics. Vague responses in conversation
destroy me. I need to know what is going on.
I need to know specifics and how to help. It is in my nature to help.
If I get annoyed with you for being vague, please forgive me.
I'm working on it.

To you it may seem odd when you hear me tell you to "please be careful." To you, you and your siblings are just playing. It's innocent, and it is, but I tell you this because of the amount of parents who have called me on 911 to request EMS because their kids were just playing when something crazy happened. I'm sorry I say it but I just want to protect you.

At the end of my work week I am so fatigued and overstimulated and traumatized that, yes, if I have to make literally one more decision I might fall entirely to pieces.

Please stop asking us "what's the worst thing you've ever heard?", because I think about it every time I close my eyes. I hear the sounds as I'm trying to fall asleep at night. Instead, ask me to tell you a funny story or a happy story. I will gladly tell you about guiding a woman through labour, to then hear the healthy cries of her baby. We are always trying to forget that "worst call", and we don't want to be forced to talk about it.

Sleep is a thing of the past. When I do, I wake up thinking I'm hearing 911 ringing. So many people tell me to "leave it at work" but that's not possible. When you hear the things we hear, it changes you. So I say "I love you" and "be careful" all too often because life is short and life is unfair. & when you're a 911 dispatcher, you have a front-row seat to all of life's unfairness.

No one leaves dispatch the same person they entered as. It doesn't matter if they are a dispatcher for 20 minutes or 20 years, it changes you. As a dispatcher, though, you get to choose if it's going to change you for the better or for the worse.

I don't want to be the person I became after taking the job that changed me. I also miss the light I used to be. The job took so much from it. I'm sorry I couldn't pull away before I had your brother and you had to see me change from the calls. I'm trying so hard to heal for you and him.

It's okay to ask for help. I wish this had been a part of the culture in my early years of dispatching. Maybe things would've been different for myself and so many coworkers struggling with how the job changes us all. I'll never be the old me prior to dispatch but I finally got the help I needed. Although my career ended because it completely destroyed my mental health - I am still healing, and will never be the old me but I'm really proud of the new me.

When I work nightshift and offer to get up early to do something before yet another 12 hour shift, please don't tell me no, that I need my sleep, and then bitterly complain that you never get to see me or we never get to do things. I offer to make sacrifices and do things when and where I can because I know shift work is hard on more than just the employee. Adding bitterness to that just makes things worse.

Every struggle... Every bad memory...
Lives inside my head, behind a smile, behind
the words of false affirmations, but one day...
one day I know the smile will fall flat and the words hollow...
and I'll have no choice but to break... but until then,
I'll keep smiling, I'll keep loving, I'll just keep thriving.

I don't get to choose the calls. I don't get to choose who yells at me or who begs me to help them. I understand that I chose the job but I don't get to choose the damage. So don't tell me that I chose this path.

When you speak to me and it seems like I am not there, I promise you that I am. I am trained to hear every detail, every tone of voice, everything. Just know that I am there. I am there, in more ways than one, but I am there, and I am listening.

The bad call you heard about? I don't want
to talk about it. I want to tuck it away
and never think about it again.

It's ingrained in me to obtain the story quickly. My patience for long, drawn-out stories are nearly non-existent. I'm working on getting better about it.

We are used to multitasking. I really
am paying attention but sometimes
I need to do more than one thing
at a time.

I'm not intentionally trying to be rude or uncaring when I'm interrupting your story or pushing you to move along in the process. My brain is wired to quickly get to the point, seeking the most important details and it is extremely hard to turn that off.

Please don't ask me what the worst thing I've ever heard was. Ask the funniest, ask the best, ask what the dumbest reason someone has called for but unless you have done the job you will never understand the worst things.

That I don't mean to be short-tempered sometimes and
grumpy. I used all my patience at work and although
I know it bothers you when I'm short or distant, I sometimes
can't help it. It doesn't mean it doesn't bother me to hurt
your feelings though, I'm trying.

Sometimes we need silence. No phones, no TV... No communication. Just silence. We are hammered all day long with so many people needing our immediate attention that we need some time to decompress.

My heart aches for all the special events I've missed, there are no do-overs...
I'd be there if I could...
I miss you...

Thank you

for what you do

Thank you for what you do

Before 2016 you didn't see much about the 911 professionals who answer the call for help. You didn't see them on social media, what little social media there was, and you didn't see many feel-good stories either. You also didn't see many stories about 911 professionals and their mental health. The only thing anyone would see would be something in the news about what dispatch had done wrong. After the #IAM911 Movement launched, it seemed like things began to change. It wasn't all because of the movement, but it certainly played a major factor in the shift of how people see what is done and what comes with answering 911 calls. It was once put that the #IAM911 Movement is the single greatest public awareness campaign showing what it is truly like to work in the profession of 911.

911 dispatch stories from all over the world were being shared. People were connecting, learning, and relating to each story. The human side of 911 pulled people in. Some stories seemed unbelievable, but they were all true. A shift was happening and now the most important person you will never see was out in the forefront of a massive movement. I was watching this all unfold over all facets of media. I even had the chance to experience it in person.

One morning I was sitting on a plane waiting to take off. I was returning after a week at a conference where I had recorded episodes of the Within the Trenches podcast, and I had presented my Imagine Listening session. Although it was early I was beaming from a great week. As I sat on the plane waiting to take off, I heard a girl's voice.

"Thank you for your service."

I picked up my head. "She couldn't be talking to me," I thought.

I looked around me and then made eye contact with the girl who had to be around 16 years old.

"Yes you," she said to me.

I was confused because I had only ever heard people say something like that to someone who is in the military. I scrambled to figure out why she was thanking me, and it hit me. It was my hat. I have a ball cap that I wear backward because of a patch that has "#IAM911" with a Thin Gold Line through it. I display it with pride, but I never thought I would experience something like this. I reached up and felt the patch with my hand.

"Yes. Thank you for what you do."

The girl smiled and I thanked her for what she said. I had to bite my lip as I began to tear up a little. I was full of emotion as she and her mother walked past me.

"What does he do," her mother asked.
"Didn't you see his hat? He's 911," the girl explained.

I sat there with tears rolling down my face. They were happy tears because I knew that things were truly changing and changing for the better. To my fellow sisters and brothers of the Thin Gold Line; you are seen. You are heard. One voice can change the world. That voice is yours. And even if you feel as though your voice can't change the world, remember this; one voice, your voice, can change your caller's world. That's the power you hold, and that's how important you are. You do a job that not many people can do, so please, hold your head up high.

Thank you for everything you do.

Ricardo Martinez II founded the #IAM911 movement in 2016, however, the curtain to the world of 9-1-1 dispatch was opened in 2010 with, "Behind the Mic - Stories from the Trenches." This was a college project where a story needed to be told as part of digital storytelling through music, narration, and still photos. Ricardo had struggled with 9-1-1 calls that he had buried, but found it therapeutic to write about them. Because people had asked over and over what his craziest call was, he decided to educate the public by sharing his stories. For this project, however, he enlisted two co-workers to share the basics, how did they get into dispatching, what was their worst call, their best call, and why they do what they do. A heartfelt shoutout goes to Whitney for her early contributions to the show, but especially to Trista and Alexis who bravely shared their stories in 2010. It is the first version of what has become the Within the Trenches podcast, the #IAM911 movement, and Imagine Listening.

To find out more or share your **#IAM911** story, visit
www.withinthetrenches.net

Printed in the USA
CPSIA information can be obtained
at www.ICGtesting.com
LVHW050532170824
788495LV00011B/27